Guangding Liu
Changchun Yang
Tianyao Hao
Xiaorong Luo

T0092689

Oil and Gas Resources in China: A Roadmap to 2050

Chinese Academy of Sciences

Guangding Liu
Changchun Yang
Tianyao Hao
Xiaorong Luo

Editors

Oil and Gas Resources in China: A Roadmap to 2050

With 17 figures

Science Press
Beijing

Springer

Editors

Guangding Liu
Institute of Geology and Geophysics, CAS
100029, Beijing, China
E-mail: gdliu@mail.igcas.ac.cn

Changchun Yang
Institute of Geology and Geophysics, CAS
100029, Beijing, China
E-mail: ccy@mail.igcas.ac.cn

Tianyao Hao
Institute of Geology and Geophysics, CAS
100029, Beijing, China
E-mail: tyhao@mail.iggcas.ac.cn

Xiaorong Luo
Institute of Geology and Geophysics, CAS
100029, Beijing, China
E-mail: luoxr@mail.iggcas.ac.cn

ISBN 978-7-03-027263-8
Science Press Beijing

ISBN 978-3-642-13903-1   e-ISBN 978-3-642-13904-8
Springer Heidelberg Dordrecht London New York

Library of Congress Control Number: 2010928539

Cover design: Frido Steinen-Broo, EStudio Calamar, Spain

Printed on acid-free paper

Springer is a part of Springer Science+Business Media (www.springer.com)

Members of the Editorial Committee and the Editorial Office

Editor-in-Chief

Yongxiang Lu

Editorial Committee

Yongxiang Lu   Chunli Bai   Erwei Shi   Xin Fang   Zhigang Li   Xiaoye Cao   Jiaofeng Pan

Research Group on Oil and Gas Resources of the Chinese Academy of Sciences

Director:

Guangding Liu   Institute of Geology and Geophysics, CAS

Vice Director:

Changchun Yang   Institute of Geology and Geophysics, CAS

Members:

Tianyao Hao   Institute of Geology and Geophysics, CAS

Xiaorong Luo   Institute of Geology and Geophysics, CAS

Bin Xia   Guangzhou Institute of Geochemistry,CAS

Xianming Xiao   Guangzhou Institute of Geochemistry,CAS

Shaoping Zhou   Bureau of Science and Technology for Resources and Environment, CAS

Ying Fan   Institute of Policy and Management, CAS

Junwei Zheng   Lanzhou Branch of the National Science Library,CAS

Foreword to the Roadmaps 2050[*]

China's modernization is viewed as a transformative revolution in the human history of modernization. As such, the Chinese Academy of Sciences (CAS) decided to give higher priority to the research on the science and technology (S&T) roadmap for priority areas in China's modernization process. What is the purpose? And why is it? Is it a must? I think those are substantial and significant questions to start things forward.

Significance of the Research on China's S&T Roadmap to 2050

We are aware that the National Mid- and Long-term S&T Plan to 2020 has already been formed after two years' hard work by a panel of over 2000 experts and scholars brought together from all over China, chaired by Premier Wen Jiabao. This clearly shows that China has already had its S&T blueprint to 2020. Then, why did CAS conduct this research on China's S&T roadmap to 2050?

In the summer of 2007 when CAS was working out its future strategic priorities for S&T development, it realized that some issues, such as energy, must be addressed with a long-term view. As a matter of fact, some strategic researches have been conducted, over the last 15 years, on energy, but mainly on how to best use of coal, how to best exploit both domestic and international oil and gas resources, and how to develop nuclear energy in a discreet way. Renewable energy was, of course, included but only as a supplementary energy. It was not yet thought as a supporting leg for future energy development. However, greenhouse gas emissions are becoming a major world concern over

---

*   It is adapted from a speech by President Yongxiang Lu at the first High-level Workshop on China's S&T Roadmap for Priority Areas to 2050, organized by the Chinese Academy of Sciences, in October, 2007.

the years, and how to address the global climate change has been on the agenda. In fact, what is really behind is the concern for energy structure, which makes us realize that fossil energy must be used cleanly and efficiently in order to reduce its impact on the environment. However, fossil energy is, pessimistically speaking, expected to be used up within about 100 years, or optimistically speaking, within about 200 years. Oil and gas resources may be among the first to be exhausted, and then coal resources follow. When this happens, human beings will have to refer to renewable energy as its major energy, while nuclear energy as a supplementary one. Under this situation, governments of the world are taking preparatory efforts in this regard, with Europe taking the lead and the USA shifting to take a more positive attitude, as evidenced in that: while fossil energy has been taken the best use of, renewable energy has been greatly developed, and the R&D of advanced nuclear energy has been reinforced with the objective of being eventually transformed into renewable energy. The process may last 50 to 100 years or so. Hence, many S&T problems may come around. In the field of basic research, for example, research will be conducted by physicists, chemists and biologists on the new generation of photovoltaic cell, dye-sensitized solar cells (DSC), high-efficient photochemical catalysis and storage, and efficient photosynthetic species, or high-efficient photosynthetic species produced by gene engineering which are free from land and water demands compared with food and oil crops, and can be grown on hillside, saline lands and semi-arid places, producing the energy that fits humanity. In the meantime, although the existing energy system is comparatively stable, future energy structure is likely to change into an unstable system. Presumably, dispersive energy system as well as higher-efficient direct current transmission and storage technology will be developed, so will be the safe and reliable control of network, and the capture, storage, transfer and use of CO_2, all of which involve S&T problems in almost all scientific disciplines. Therefore, it is natural that energy problems may bring out both basic and applied research, and may eventually lead to comprehensive structural changes. And this may last for 50 to 100 years or so. Taking the nuclear energy as an example, it usually takes about 20 years or more from its initial plan to key technology breakthroughs, so does the subsequent massive application and commercialization. If we lose the opportunity to make foresighted arrangements, we will be lagging far behind in the future. France has already worked out the roadmap to 2040 and 2050 respectively for the development of the 3rd and 4th generation of nuclear fission reactors, while China has not yet taken any serious actions. Under this circumstance, it is now time for CAS to take the issue seriously, for the sake of national interests, and to start conducting a foresighted research in this regard.

This strategic research covers over some dozens of areas with a long-term view. Taking agriculture as an example, our concern used to be limited only to the increased production of high-quality food grains and agricultural by-products. However, in the future, the main concern will definitely be given to the water-saving and ecological agriculture. As China is vast in territory, diversified technologies in this regard are the appropriate solutions. Animal husbandry has been used by developed countries, such as Japan and Denmark, to make bioreactor and pesticide as well. Plants have been used by Japan to make bioreactors which are safer and cost-effective than that made from animals. Potato, strawberry, tomato and the like have been bred in germ-free greenhouses, and value-added products have been made through gene transplantation technology. Agriculture in China must not only address the food demands from its one billions-plus population, but also take into consideration of the value-added agriculture by-products and the high-tech development of agriculture as well. Agriculture in the future is expected to bring out some energies and fuels needed by both industry and man's livelihood as well. Some developed countries have taken an earlier start to conduct foresighted research in this regard, while we have not yet taken sufficient consideration.

Population is another problem. It will be most likely that China's population will not drop to about 1 billion until the end of this century, given that the past mistakes of China's population policy be rectified. But the subsequent problem of ageing could only be sorted out until the next century. The current population and health policies face many challenges, such as, how to ensure that the 1.3 to 1.5 billion people enjoy fair and basic public healthcare; the necessity to develop advanced and public healthcare and treatment technologies; and the change of research priority to chronic diseases from infectious diseases, as developed countries have already started research in this regard under the increasing social and environmental change. There are many such research problems yet to be sorted out by starting from the basic research, and subsequent policies within the next 50 years are in need to be worked out.

Space and oceans provide humanity with important resources for future development. In terms of space research, the well-known Manned Spacecraft Program and China's Lunar Exploration Program will last for 20 or 25 years. But what will be the whole plan for China's space technology? What is the objective? Will it just follow the suit of developed countries? It is worth doing serious study in this regard. The present spacecraft is mainly sent into space with chemical fuel propellant rocket. Will this traditional propellant still be used in future deep space exploration? Or other new technologies such as electrical propellant, nuclear energy propellant, and solar sail technologies be developed? We haven't yet done any strategic research over these issues, not even worked out any plans. The ocean is abundant in mineral resources, oil and gas, natural gas hydrate, biological resources, energy and photo-free biological evolution, which may arise our scientific interests. At present, many countries have worked out new strategic marine plans. Russia, Canada, the USA, Sweden and Norway have centered their contention upon the North Pole, an area of strategic significance. For this, however, we have only limited plans.

The national and public security develops with time, and covers both conventional and non-conventional security. Conventional security threats only refer to foreign invasion and warfare, while, the present security threat may come out from any of the natural, man-made, external, interior, ecological, environmental, and the emerging networking (including both real and virtual) factors. The conflicts out of these must be analyzed from the perspective of human civilization, and be sorted out in a scientific manner. Efforts must be made to root out the cause of the threats, while human life must be treasured at any time.

In general, it is necessary to conduct this strategic research in view of the future development of China and mankind as well. The past 250 years' industrialization has resulted in the modernization and better-off life of less than 1 billion people, predominantly in Europe, North America, Japan and Singapore. The next 50 years' modernization drive will definitely lead to a better-off life for 2–3 billion people, including over 1 billion Chinese, doubling or tripling the economic increase over that of the past 250 years, which will, on the one hand, bring vigor and vitality to the world, and, on the other hand, inevitably challenge the limited resources and eco-environment on the earth. New development mode must be shaped so that everyone on the earth will be able to enjoy fairly the achievements of modern civilization. Achieving this requires us, in the process of China's modernization, to have a foresighted overview on the future development of world science and human civilization, and on how science and technology could serve the modernization drive. S&T roadmap for priority areas to 2050 must be worked out, and solutions to core science problems and key technology problems must be straightened out, which will eventually provide consultations for the nation's S&T decision-making.

Possibility of Working out China's S&T Roadmap to 2050

Some people held the view that science is hard to be predicted as it happens unexpectedly and mainly comes out of scientists' innovative thinking, while, technology might be predicted but at the maximum of 15 years. In my view, however, S&T foresight in some areas seems feasible. For instance, with the exhaustion of fossil energy, some smart people may think of transforming solar energy into energy-intensive biomass through improved high-efficient solar thin-film materials and devices, or even developing new substitute. As is driven by huge demands, many investments will go to this emerging area. It is, therefore, able to predict that, in the next 50 years, some breakthroughs will undoubtedly be made in the areas of renewable energy and nuclear energy as well. In terms of solar energy, for example, the improvement of photoelectric conversion efficiency and photothermal conversion efficiency will be the focus. Of course, the concrete technological solutions may be varied, for example, by changing the morphology of the surface of solar cells and through the reflection, the entire spectrum can be absorbed more efficiently; by developing multi-layer functional thin-films for transmission and absorption; or by introducing of nanotechnology and quantum control technology, etc. Quantum control research used to limit mainly to the solution to information functional materials. This is surely too narrow. In the future, this research is expected to be extended to the energy issue or energy-based basic research in cutting-edge areas.

In terms of computing science, we must be confident to forecast its future development instead of simply following suit as we used to. This is a possibility rather than wild fancies. Information scientists, physicists and biologists could be engaged in the forward-looking research. In 2007, the Nobel Physics Prize was awarded to the discovery of colossal magneto-resistance, which was, however, made some 20 years ago. Today, this technology has already been applied to hard disk store. Our conclusion made, at this stage, is that: it is possible to make long-term and unconventional S&T predictions, and so is it to work out China's S&T roadmap in view of long-term strategies, for example, by 2020 as the first step, by 2030 or 2035 as the second step, and by 2050 as the maximum.

This possibility may also apply to other areas of research. The point is to emancipate the mind and respect objective laws rather than indulging in wild fancies. We attribute our success today to the guidelines of emancipating the mind and seeking the truth from the facts set by the Third Plenary Session of the 11th Central Committee of the Communist Party of China in 1979. We must break the conventional barriers and find a way of development fitting into China's reality. The history of science tells us that discoveries and breakthroughs could only be made when you open up your mind, break the conventional barriers, and make foresighted plans. Top-down guidance on research with increased financial support and involvement of a wider range of talented scientists is not in conflict with demand-driven research and free discovery of science as well.

Necessity of CAS Research on China's S&T Roadmap to 2050

Why does CAS launch this research? As is known, CAS is the nation's highest academic institution in natural sciences. It targets at making basic, forward-looking and strategic research and playing a leading role in China's science. As such, how can it achieve this if without a foresighted view on science and technology? From the perspective of CAS, it is obligatory to think, with a global view, about what to do after the 3rd Phase of the Knowledge Innovation Program (KIP). Shall we follow the way as it used to? Or shall we, with a view of national interests, present our in-depth insights into different research disciplines, and make efforts to reform the organizational structure and system, so that the innovation capability of CAS and the nation's science and technology mission will be raised to a new height? Clearly, the latter is more positive. World science and technology develops at a lightening speed. As global economy grows, we are aware that we will be lagging far behind if without making progress, and will lose the opportunity if without making foresighted plans. S&T innovation requires us to make joint efforts, break the conventional barriers and emancipate the mind. This is also what we need for further development.

The roadmap must be targeted at the national level so that the strategic research reports will form an important part of the national long-term program. CAS may not be able to fulfill all the objectives in the reports. However, it can select what is able to do and make foresighted plans, which will eventually help shape the post-2010 research priorities of CAS and the guidelines for its future reform.

Once the long-term roadmap and its objectives are identified, system mechanism, human resources, funding and allocation should be ensured for full implementation. We will make further studies to figure out: What will happen to world innovation system within the next 30 to 50 years? Will universities, research institutions and enterprises still be included in the system? Will research institutes become grid structure? When the cutting-edge research combines basic science and high-tech and the transformative research integrates the cutting-edge research with industrialization, will that be the research trend in some disciplines? What will be the changes for personnel structure, motivation mechanism and upgrading mechanism within the innovation system? Will there be any changes for the input and structure of innovation resources? If we could have a clear mind of all the questions, make foresighted plans and then dare to try out in relevant CAS institutes, we will be able to pave a way for a more competitive and smooth development.

Social changes are without limit, so are the development of science and technology, and innovation system and management as well. CAS must keep moving ahead to make foresighted plans not only for science and technology, but also for its organizational structure, human resources, management modes, and resource structures. By doing so, CAS will keep standing at the forefront of science and playing a leading role in the national innovation system, and even, frankly speaking, taking the lead in some research disciplines in the world. This is, in fact, our purpose of conducting the strategic research on China's S&T roadmap.

Prof. Dr.-Ing. Yongxiang Lu
President of the Chinese Academy of Sciences

Oil and Gas Resources in China: A Roadmap to 2050

Preface to the Roadmaps 2050

CAS is the nation's think tank for science. Its major responsibility is to provide S&T consultations for the nation's decision-makings and to take the lead in the nation's S&T development.

In July, 2007, President Yongxiang Lu made the following remarks: "In order to carry out the Scientific Outlook of Development through innovation, further strategic research should be done to lay out a S&T roadmap for the next 20–30 years and key S&T innovation disciplines. And relevant workshops should be organized with the participation of scientists both within CAS and outside to further discuss the research priorities and objectives. We should no longer confine ourselves to the free discovery of science, the quantity and quality of scientific papers, nor should we satisfy ourselves simply with the Principal Investigators system of research. Research should be conducted to address the needs of both the nation and society, in particular, the continued growth of economy and national competitiveness, the development of social harmony, and the sustainability between man and nature. "

According to the Executive Management Committee of CAS in July, 2007, CAS strategic research on S&T roadmap for future development should be conducted to orchestrate the needs of both the nation and society, and target at the three objectives: the growth of economy and national competitiveness, the development of social harmony, and the sustainability between man and nature.

In August, 2007, President Yongxiang Lu further put it: "Strategic research requires a forward-looking view over the world, China, and science & technology in 2050. Firstly, in terms of the world in 2050, we should be able to study the perspectives of economy, society, national security, eco-environment, and science & technology, specifically in such scientific disciplines as energy, resources, population, health, information, security, eco-environment, space and oceans. And we should be aware of where the opportunities and challenges lie. Secondly, in terms of China's economy and society in 2050, we should take into consideration of factors like: objectives, methods, and scientific supports needed for economic structure, social development, energy structure, population and health, eco-environment, national security and innovation capability. Thirdly, in terms of the guidance of Scientific Outlook of Development on science and technology, it emphasizes the people's interests and development, science and technology, science and economy, science and society, science and eco-environment, science and culture, innovation and collaborative development. Fourthly, in terms of the supporting role of research in scientific development, this includes how to optimize the economic structure and boost economy, agricultural development, energy structure, resource conservation, recycling economy, knowledge-based society, harmonious coexistence between man and nature, balance of regional development, social harmony, national security, and international cooperation. Based on these, the role of CAS will be further identified."

Subsequently, CAS launched its strategic research on the roadmap for priority areas to 2050, which comes into eighteen categories including: energy, water resources, mineral resources, marine resources, oil and gas, population and health, agriculture, eco-environment, biomass resources, regional development, space, information, advanced manufacturing, advanced materials, nano-science, big science facilities, cross-disciplinary and frontier research, and national and public security. Over 300 CAS experts in science, technology, management and documentation & information, including about 60 CAS members, from over 80 CAS institutes joined this research.

Over one year's hard work, substantial progress has been made in each research group of the scientific disciplines. The strategic demands on priority areas in China's modernization drive to 2050 have been strengthened out; some core science problems and key technology problems been set forth; a relevant S&T roadmap been worked out based on China's reality; and eventually the strategic reports on China's S&T roadmap for eighteen priority areas to 2050 been formed. Under the circumstance, both the Editorial Committee and Writing Group, chaired by President Yongxiang Lu, have finalized the general report. The research reports are to be published in the form of CAS strategic research serial reports, entitled *Science and Technology Roadmap to China 2050: Strategic Reports of the Chinese Academy of Sciences*.

The unique feature of this strategic research is its use of S&T roadmap approach. S&T roadmap differs from the commonly used planning and technology foresight in that it includes science and technology needed for the future, the roadmap to reach the objectives, description of environmental changes, research needs, technology trends, and innovation and technology development. Scientific planning in the form of roadmap will have a clearer scientific objective, form closer links with the market, projects selected be more interactive and systematic, the solutions to the objective be defined, and the plan be more feasible. In addition, by drawing from both the foreign experience on roadmap research and domestic experience on strategic planning, we have formed our own ways of making S&T roadmap in priority areas as follows:

(1) Establishment of organization mechanism for strategic research on S&T roadmap for priority areas

The Editorial Committee is set up with the head of President Yongxiang Lu and the involvement of Chunli Bai, Erwei Shi, Xin Fang, Zhigang Li, Xiaoye Cao and Jiaofeng Pan. And the Writing Group was organized to take responsibility of the research and writing of the general report. CAS Bureau of Planning and Strategy, as the executive unit, coordinates the research, selects the scholars, identifies concrete steps and task requirements, sets forth research approaches, and organizes workshops and independent peer reviews of the research, in order to ensure the smooth progress of the strategic research on the S&T roadmap for priority areas.

(2) Setting up principles for the S&T roadmap for priority areas

The framework of roadmap research should be targeted at the national level, and divided into three steps as immediate-term (by 2020), mid-term (by 2030) and long-term (by 2050). It should cover the description of job requirements, objectives, specific tasks, research approaches, and highlight core science problems and key technology problems, which must be, in general, directional, strategic and feasible.

(3) Selection of expertise for strategic research on the S&T roadmap

Scholars in science policy, management, information and documentation, and chief scientists of the middle-aged and the young should be selected to form a special research group. The head of the group should be an outstanding scientist with a strategic vision, strong sense of responsibility and coordinative capability. In order to steer the research direction, chief scientists should be selected as the core members of the group to ensure that the strategic research in priority areas be based on the cutting-edge and frontier research. Information and documentation scholars should be engaged in each research group to guarantee the efficiency and systematization of the research through data collection and analysis. Science policy scholars should focus on the strategic demands and their feasibility.

(4) Organization of regular workshops at different levels

Workshops should be held as a leverage to identify concrete research steps and ensure its smooth progress. Five workshops have been organized consecutively in the following forms:

High-level Workshop on S&T Strategies. Three workshops on S&T strategies have been organized in October, 2007, December, 2007, and June, 2008, respectively, with the participation of research group heads in eighteen priority areas, chief scholars, and relevant top CAS management members. Information has been exchanged, and consensus been reached to ensure research directions. During the workshops, President Yongxiang Lu pinpointed the significance, necessity and possibility of the roadmap research, and commented on the work of each research groups, thus pushing the research forward.

Special workshops. The Editorial Committee invited science policy scholars to the special workshops to discuss the eight basic and strategic systems for China's socio-economic development. Perspectives on China's science-driven modernization to 2050 and characteristics and objectives of the eight systems have been outlined, and twenty-two strategic S&T problems affecting the modernization have been figured out.

Research group workshops. Each research group was further divided into different research teams based on different disciplines. Group discussions, team discussions and cross-team discussions were organized for further research, occasionally with the involvement of related scholars in special topic discussions. Research group workshops have been held some 70 times.

Cross-group workshops. Cross-group and cross-disciplinary workshops were organized, with the initiation by relative research groups and coordination by Bureau of Planning and Strategies, to coordinate the research in relative disciplines.

Professional workshops. These workshops were held to have the suggestions and advices of both domestic and international professionals over the development and strategies in related disciplines.

(5) Establishment of a peer review mechanism for the roadmap research

To ensure the quality of research reports and enhance coordination among different disciplines, a workshop on the peer review of strategic research on the S&T roadmap was organized by CAS Bureau of Planning and Strategy, in November, 2008, bringing together of about 30 peer review experts and 50 research group scholars. The review was made in four different categories, namely, resources and environment, strategic high-technology, bio-science & technology, and basic research. Experts listened to the reports of different research groups, commented on the general structure, what's new and existing problems, and presented their suggestions and advices. The outcomes were put in the written forms and returned to the research groups for further revisions.

(6) Establishment of a sustained mechanism for the roadmap research

To cope with the rapid change of world science and technology and national demands, a roadmap is, by nature, in need of sustained study, and should be revised once in every 3–5 years. Therefore, a panel of science policy scholars should be formed to keep a constant watch on the priority areas and key S&T problems for the nation's long-term benefits and make further study in this regard. And hopefully, more science policy scholars will be trained out of the research process.

The serial reports by CAS have their contents firmly based on China's reality while keeping the future in view. The work is a crystallization of the scholars' wisdom, written in a careful and scrupulous manner. Herewith, our sincere gratitude goes to all the scholars engaged in the research, consultation and review. It is their joint efforts and hard work that help to enable the serial reports to be published for the public within only one year.

To precisely predict the future is extremely challenging. This strategic research covered a wide range of areas and time, and adopted new research approaches. As such, the serial reports may have its deficiency due to the limit in knowledge and assessment. We, therefore, welcome timely advice and enlightening remarks from a much wider circle of scholars around the world.

The publication of the serial reports is a new start instead of the end of the strategic research. With this, we will further our research in this regard, duly release the research results, and have the roadmap revised every five years, in an effort to provide consultations to the state decision-makers in science, and give suggestions to science policy departments, research institutions, enterprises, and universities for their S&T policy-making. Raising the public awareness of science and technology is of great significance for China's modernization.

Writing Group of the General Report

February, 2009

Preface

In Oct. 2007, Chinese Academy of Sciences launched the campaign of the strategic research on development roadmap in key scientific and technological fields to 2050 to carry out study on the development roadmap in 18 key scientific and technological fields, the science and technology on oil and gas resources is one of the 18 key fields.

According to the general arrangement and requirements of the strategic research on development roadmap in key scientific and technological fields in China made by Chinese Academy of Sciences, the oil and gas resource group, based on the current oil and gas resource situation and environment in China, and the national energy strategic principle on giving priority to domestic exploration and development, rational importing of oil and gas according to market rule, trying to build the energy saving society and developing renewable clean energy, made scientific researches on the trend of the development strategy in oil and gas resources in other countries, particularly the developed countries in the world. In addition, based on the special geology background in oil and gas exploration and development as well as the oil and gas resources requirement for future economic and social development in China, the oil and gas group also made a research of the development and strategic objective and the near term objectives, medium term objectives and long term objectives on oil and gas resources in China to 2050.

To make track for the forefront of the development in oil and gas technology in the world to grasp the main development trends in science and technology and/or the major technologic problems to be tackled at all stages, based on which, we should solicit opinions from the relevant experts and put forward the program for scientific and technological development in oil and gas for the next 50 years in China, formulate the strategic roadmap of the oil and gas exploration and development technology in China, and propose the policies in terms of systems, resources and talents needed for the development objective.

The researches are as the follows:

1) Pay close attention to the future development strategies and development trends on oil and gas resources in developed countries

We shall further understand the general demand, proportion and exploration model needed for oil and gas resources in terms of energy, resource and environment in international community in the next half century based on the current technologies adopted by the developed countries and large oil companies as well as their strategic deployments and technological development trends.

2) Make a research of the development and strategic objectives and the near term objectives, medium term objectives and long term objectives on oil and gas resources in China to 2050

Take the national oil and gas resource occurrence environment and condition and the requirement of the national economic development in terms of energy, resource and environment into comprehensive consideration, and refer to the development trend on oil and gas resource in international communities so as to analyze the overall objectives and interim objectives at different stages for the development strategies in terms of the national oil and gas resources in China.

3) Grasp the main development trends in science and technology and/or the major technologic problems to be tackled at all stages

According to the analysis of the leading technologies adopted for the current international oil and gas exploration and development trends, we should find out the major technological issues suitable to our practical geologic condition and social environment.

4) Formulate the strategic roadmap for the national oil and gas exploration and development technologies

Put forward the program for the technological development on oil and gas resources in the next 50 years based on the preparatory work and the above investigation work and present the strategic roadmap for the oil and gas exploration and development technologies suitable to the territorial and social characteristics in China.

5) Propose the rational and feasible policies on system, resource and talent needed to achieve the development objective

Put forward the opinion and proposal as well as practicable policy in terms of the restructuring, social resource utilization, talent education and how to bring efficacy into play according to the above knowledge, together with the overall objectives of the socialist market economy.

Collect and collate the information on development strategies in terms of oil and gas resources in countries, particularly the developed countries in the world according to the knowledge obtained from the above research together with the professional advantages of the research group members and cooperation between the members in the research group so as to research and analyze their development trend and then prepare reports respectively, which shall be finalized by the group leader, during which period, many panel discussions have been held within the group members. The group also participated in the seminars for development roadmap in key scientific and technologic fields sponsored by Bureau of Planning & Strategy of Chinese Academy of Sciences for three times so as to give audience to the supplementary comments in every direction. After many times of modifications, the document has been finalized.

The document is composed of 5 chapters: Chapter 1 analyzes the demand and situation in terms of national oil and gas resources in China; Chapter 2 makes a research of the main fields and problems in terms of oil and gas exploration and development; Chapter 3 predicts the overall objectives of the scientific and technological development in terms of the oil and gas resources in China; Chapter 4 presents the primary scientific research orientation in terms of oil and gas resources by Chinese Academy of Sciences; Chapter 5 proposes the measures on security system establishment for the scientific and technological development in terms of oil and gas resources in China.

Contents

Abstract

Oil and gas resources are the important strategic resources directly concerned with socioeconomic development and national energy security. It is inevitable for China to ensure continuous supply of oil and gas resources for modernization drive. Since China became a net importer for oil in 1993, its crude oil consumption has been increasing by 5.77 annually and China has become the second largest oil consumption country in the world, under the economic globalization, the participating countries compete and share the global oil and gas resources by importing oil and gas to cover the shortage of oil and gas, which has been greatly promoting our national economic development. But the strategy for oil import is limited by many circumstances, as a big power, China has to adhere to the principle of giving priority to development of the domestic oil and gas resources to maintain and increase its self-sufficiency rate as the economic growth so as to maintain independence in terms of politics and diplomacy and safeguard national interests, based on which, we should further exploit overseas market of oil and gas resources to take opportunity to import the oil and gas from overseas to protect domestic oil and gas resources and maintain high speed growth of national economy and maximize the national interests.

From the above explanation of the situation and environment of the oil and gas resources in face of China, we have to stick to the energy strategy principle of "giving priority to exploration and development of domestic resources, rational importing oil and gas according to market rule, trying to build energy saving society and developing renewable clean energy" for the several decades in future, which is also the program to guide the technological development in terms of oil and gas resources. Thereby, given the special geologic background in China and the forefront of the global technological development in terms of oil and gas resources, the strategic roadmap for our national oil and gas resource development technology is formulated.

Near term (about 2025), to improve the theoretical system for the basin formation and evolution as well as oil and gas deposit and establish the theory on marine facies carbonate rock oil and gas deposits preliminarily, with focus on development of the geophysical exploration technology on complex topography, and create the theory and technology for oil and gas exploration and development in Mesozoic carbonate rock layer system and enhance the exploration efficiency of carbonate rock layer system; develop high-resolution seismic work and explanation technology, further research on the forming mechanism of the hydrocarbon reservoir in low permeability, fissure and unconventional reservoir beds and improve the exploration precision of the high quality reservoir space; make a research of the distribution mechanism of the remaining oil, develop the bio-technology for oil production, adopt rational secondary recovery and tertiary recovery technology, enhance oil recovery, introduce and develop the deep drilling equipment, carry on exploration of the oil and gas in deep layer, deep sea, and semi-deep sea, research the corresponding exploration technology, confirm the existence of the methane hydrate deposit in sea area and Qinghai-Tibet Plateau permafrost zone in China, research the mechanism and environment of their formation and conservation, analyze the drilling and gas desorption technology applicable to our coalbed gas reservoir; explore the high efficient refining method for asphalite and oil shale, expand overseas exploration and development scope, pay attention to the cooperation in terms of the international oil and gas exploration in Arctic Ocean Circle.

Medium term (about 2035), to grasp the genetic type of hydrocarbon and oil and gas distribution regularity of the petroliferous basin in substance in China, establish the perfect forming theory on basin, hydrocarbon and reservoir in deep water and ultra-deep basin by and large, make a breakthrough in deep layer and ultra-deep layer oil and gas drilling and development technologies, greatly enhance the oil and gas recovery efficiency by leaps and bounds, focus on the exploration and development technologies for unconventional natural gas reservoir (such as shale gas, non-biological methane gas, deep layer natural gas etc.), develop the independent technology for oil and gas exploration and development in deep sea and semi-deep sea, introduce, improve and explore the technology applicable to our national geological conditions for exploration and development of the methane hydrate deposit, improve the deep layer drilling and exploration technology, enhance the deep layer oil and gas exploration performance and efficiency and participate in the oil and gas exploration and development in Arctic Ocean Circle and other common regions in the world.

Long term (about 2050), to carry on the oil and gas exploration and development of the deep water to ultra-deep water basin, greatly enhance the deep layer oil and gas exploration performance and recovery efficiency, participate in the oil and gas exploration and development of the Arctic Ocean Circle and other common regions. Complete the exploration and development technology system of the methane hydrate in surrounding sea area by and large, establish a series of natural gas exploration and development technologies applicable to complex geological conditions and the deep drilling and production technology suitable to the geological characteristic of the basins in China, research the technologies for reformation of the oil and gas field reservoirs having been subjected to oil and gas development for three times and the technologies for Re-aggregation of the oil and gas, thereof, establish a series of microbial enhanced oil recovery technologies.

According to the arrangement of the academic disciplines of Chinese

Academy of Sciences and the development trend in future, the leading scientific research orientations in oil and gas resources in future: continentalization, drift, matching and reformation and their restraint on our marine facies oil and gas resources in China's mainland; research on the oil and gas resources in Pre-Cenozoic marine facies relict basin; impact of the global climate evolution on hydrocarbon source rock since Mesozoic-Cenozoic; interaction between the basin deep fluid and rock together with the high quality reservoir forming mechanism; geophysical response characteristic and high precision forecasting technique of oil and gas reservoir; technology to enhance oil recovery; high density, wireless transmission earthquake acquisition system based on MEMS (Micro Electro Mechanical Systems) sensor; mass data processing and large scale hydrocarbon reservoir numerical simulation concurrent computing.

1 Requirement and Present Situation of the Oil and Gas Resources in China

Oil and natural gas (hereinafter referred to as oil and gas) resource, as a kind of necessary fundamental energy resource and chemical raw material, plays an important role in national economy and human life and is an important strategic resource related to socioeconomic development and national energy security.

Given the rapid development of the global economy and constant enhancement of life energy requirement, the demand for oil and gas has been increasing, although the global oil and gas output has been steadily growing in recent years. Its growth rate still can not satisfy the demand for oil and gas as a result of the swift growth in global economy, and the contradiction between supply and demand shall inevitably raise the oil price in the world. Since 2006, the global oil price has been on the rise, particularly, it is the first time for the global oil price exceeded US$ 100 per barrel on Jan. 2, 2008, and it has been up to the US$ 147.2 per barrel (July 11, 2008) [1] being all time record, which brings about great pressure on the economic development of the big powers in terms of oil and gas consumption. Although the oil price has been decreasing as a result of global financial crisis since september 2008, and the global economic recovery in 2009 has gradually increased the oil price to US$ 70 per barrel. In the long run, as a non-renewable strategic resource, the international oil market shall be seller's market, it is inevitable that the global oil price shall maintain at high level. As early as 1998, when the global oil price was in downturn, C. J. Campbell and J. H. Laherrère [2] published an article on *Scientific American* drawing the conclusion that the global cheap oil times shall be inevitably terminated.

It is the most important task for most countries in the world to make a search for and take advantage of the limited oil and gas resources and guarantee the rapid and stable development of the national economy for quite a long time in the future. For every country in the world, main approaches for them to safeguard their oil and gas security contain two aspects: one approach is based on domestic oil and gas resources, namely to intensify exploration and development of domestic oil and gas resources; another approach is to diversify the overseas oil and gas resources through international cooperation and the exploration and development of the domestic oil and gas resources are restricted by the domestic oil and gas resource situation as well as exploration and development technology, while taking the advantage of overseas oil and gas resources are also restricted by varying factors. There are two approaches for countries to utilize overseas oil and gas resources, one approach is to purchase oil and gas directly from the international oil and gas market, another approach is to invest and participate in overseas oil and gas exploration and development to share the oil and gas with overseas developers. Regarding direct procurement of the oil and gas from overseas market, although there is no need to participate in the oil and gas production of the exporting countries, the complicated international political relations and fluctuating oil and gas price together with the competition between the oil and gas consumption countries on the international oil and gas market make the established energy supply chain become fragile as a result of certain political interests and threaten the national energy security, while the direct investment in overseas oil and gas field to share the oil and gas resources shall be more stable in terms of oil and gas supply compared with direct procurement of oil and gas from international market, but it still be affected by the oil and gas exploration and development technology, the investment policy as well as social and political stability in the country in which an investment is to be made. Thereby, the top priority is to explore and develop domestic oil and gas resources and further safeguard oil and gas supply capability so as to ensure the national oil and gas supply.

1.1 Oil and gas resource demand

With the rapid development of the national economy, the oil and gas demand is steadily increasing, although the oil and gas output in China has been increasing in recent years, and it still can not satisfy the oil and gas demand for rapid national development.

China became a oil net importer in 1993, since then, its crude oil consumption has been increasing by 5.77% annually, while its domestic crude oil supply growth speed is only 1.67%. China has become the second largest crude oil consumption country and the third largest oil importing country next only to the United States and Japan.

China imported 145 million t of crude oil (18.36 million t higher than the figure in 2005) in 2006, equivalent to 400,000 t of crude oil import per day, spending more than US$ 66.4 billion; at the same time, China's import volume for product oil is 36.38 million t , spending US$15.551 billion. Oil and steel have become the import items spending much of the national foreign currency.

Particularly, in recent years, the oil import volume of China has been greatly increasing, which reflects the national oil consumption trend and the great disparity between oil consumption and production capacity is worrisome (Fig. 1-1): in 2005, China imported 136 million t of crude oil, with 42.9% of the rate of dependency on overseas resources; although China's crude oil output in 2006 was increased to 185 million t , its rate of dependency on overseas oil was increased to 47.4%; China's crude oil output was about 185 million t in 2007, but its oil import volume was up to 163 million t , and its rate of dependency on overseas oil was near to 50%. These increasing figures show that China exceeded the figures namely that the aggregate demand for oil in 2010 shall be up to 320 million t and that the rate of dependency on overseas oil shall be up to 44%, which are estimated by *Strategic Research Report on Sustainable Development of Oil and Gas Resources of China*[3] prepared by Chinese Academy of Engineering.

Rate of Dependency on Overseas Oil

Percentage of the crude oil import volume to the crude oil consumption.

Strategic Research Report on Sustainable Development of Oil and Gas Resources of China

The research subject "China sustainable development oil and gas resources strategy" was launched in May, 2003, the research subject group composed of 31 academicians and 120 experts and scholars organized by Chinese Academy of Engineering, and the Advisory Committee of Research Subject composed of 23 academicians and experts from Chinese Academy of Sciences and Chinese Academy of Engineering as well as major oil companies, which carried out in-depth investigation & research and the cross-disciplinary, cross-department and cross-industry demonstration, after more than one year of systematic research, completed the Strategic Research Report on Sustainable Development of Oil and Gas Resources of China.

Strategic Research Report on Sustainable Development of Oil and Gas Resources of China[3] has done a research on the supply and demand strategy of national oil and gas resources, domestic oil and gas resource development strategy, development and import strategy on overseas oil and gas resources, fuel-efficient and alternative fuel, petrochemical industry development strategy, oil and gas security and reserve strategy, regulation and policy on oil and gas, conducted comprehensive analysis on the trend between supply and demand in terms of national oil and gas resources, put forward the total strategy, guideline policy and measure for the sustainable development of the national oil and gas resources.

These data confirm the demand for oil and gas in 2030 in China forecasted by International Energy Agency (IEA) and Energy Information Administration (Fig.1-1, Table 1-1); up to 2030, the demand for oil and gas in 2030 is about 800 (EIA) million t of crude oil and 200 billion m^3. The detailed consumption shall be dependent on many factors, including the global production capacity for oil and gas and the corresponding oil price as main factors (Table 1-1).

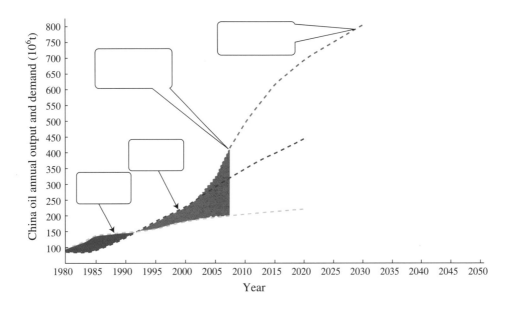

Fig.1-1　Comparison between annual output and demand in oil and gas in China
(according to the documents assembly of IEA[4],2008 ; EIA[5],2008)

Table 1-1　Oil and gas demand in 2030 in China forecasted by international organizations
(according to the documents assembly of IEA[4], 2008; EIA[5], 2008)

Organizations	EIA	IEA				
Estimation	Reference	Reference	High growth	Low growth	High oil price	Low oil price
Oil (billion t)	0.808	0.802	0.894	0.715	0.695	0.925
Natural gas (billion m³)	236	198.1	215.1	181.1	206.6	189.6

International Energy Agency

International Energy Agency, hereinafter referred to as IEA [4,6-8], is an intergovernmental organization for economic cooperation between oil consumption countries to carry on international energy programme, it was established according to the requirement of Organization for Economic Cooperation and Development in November 1974 and head quartered in Paris. At present, the member states of IEA: Australia, Austria, Belgium, Canada, Czech Republic, Danmark, Finland, France, Germany, Greece, Hungary, Ireland, Italy, Japan, Republic of Korea, Luxembourg, Netherlands, New Zealand, Norway, Poland, Portugal, Slovak Republic, Spain, Sweden, Switzerland, Turkey, United Kingdom, and United States. In addition, The European Commission also participates in the work of IEA.

The fundamental tenet of IEA: to coordinate the energy policies between member states; to increase the self-sufficiency capacity in oil supply; to coordinately take the oil demand-saving measures; to strengthen the long-term cooperation to reduce the dependence on oil import; to provide the oil market information; to draft the oil consumption plans; to share oil based on the plans in oil shortage; to facilitate the relationships between the oil manufacturer countries and the other oil consumer countries etc.

Capital work of IEA: to implement the "Urgent Plan of Oil Share" among the members in the oil shortage, which includes sharing the oil inventory among the member countries according to the agreements, limiting the crude oil consumption, dumping oil stock and etc.; to require the member countries to hold the oil inventory in a certain quantity, i.e. the oil stock which is not less than 90-day oil import; formulate the long-term cooperation programs and strengthen the energy supply security; to facilitate the stabilization of the global energy market; to develop the cooperation in energy conservation; to accelerate the progress of the alternative energy; to launch the research and development of the new energy technologies; to reform the countries' measures of energy supply on legislation and administration; to formulate the oil market information and negotiation systems to stabilize the oil market and provide promising future and enhance the oil manufacturer countries and consumer countries; to carry out the corresponding actions to the influence of the energy and environment and study the clean fuel; to regularly forecast the world energy prospects for the countries' reference.

Oil and combustible gas are the major energy consumed by human being. The global fossil energy consumption which is derived from oil, combustible gas and coal, has been keeping increasing in the recent 30 years. Other kinds of energy, such as water power, nuclear energy, bioenergy, wind and solar energy, only account for a small proportion in the total energy. Even developing with the most optimistic speed, it is not possible to replace the fossil energy and dominate (Fig.1-2) at least before the year of 2050. Therefore, nowadays and during a quite long period from now on, the oil and gas still occupy the leading position in the energy consumption demand. The forecast of IEA *World Energy Outlook* 2008[4] for the world oil and gas demand shows that the global oil need in 2030 would rise to about 5,300 Mtoe (million t oil equivalent); the natural gas need to over 3,800 Mtoe; and the coal need to near 5,000 Mtoe. Meanwhile, the IEA report also presents that the development and utilization of the global natural gas liquid would continually grow in the coming 20 years; but the oil consumption would still dominate the primary energy consumption and increase over 25% on the 2006 base, which would be much more than the total proportion of the nuclear energy, non-hydrocarbon, water and renewal resources. This situation is obviously appearing in the developing countries.

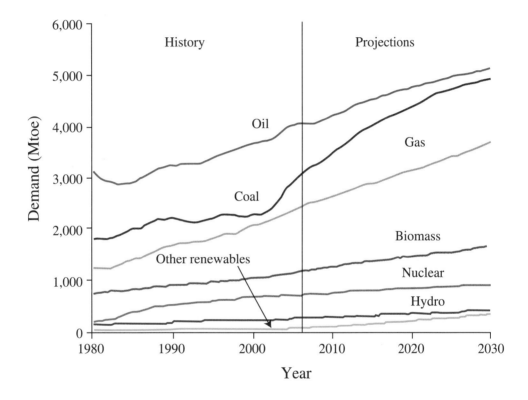

Fig. 1-2   Based on global 1980–2006 primary energy consumption estimate to 2030 demand trend [4]

In the near 30 years, the rapid development of China's economy has made its energy consumption total far exceeds the world average, especially its higher demand for oil (Fig. 1-3) and natural gas with their ever-increasing proportion in the energy consumption. The growth of a nation's GDP per capita has close connection with the oil and gas consumption. As the statistics (shown in Fig. 1-4) of the average annual oil consumption per capita and GDP per capita in the countries' economic development processes, the development in China is basically normal. Therefore, our oil and gas demand would increase at least 20 million t every year till 2025 and then slower down (Fig. 1-1 and Fig. 1-3). The oil shortage has been regarded as the huge "bottleneck" which restrains the 21st-century China's economic development.

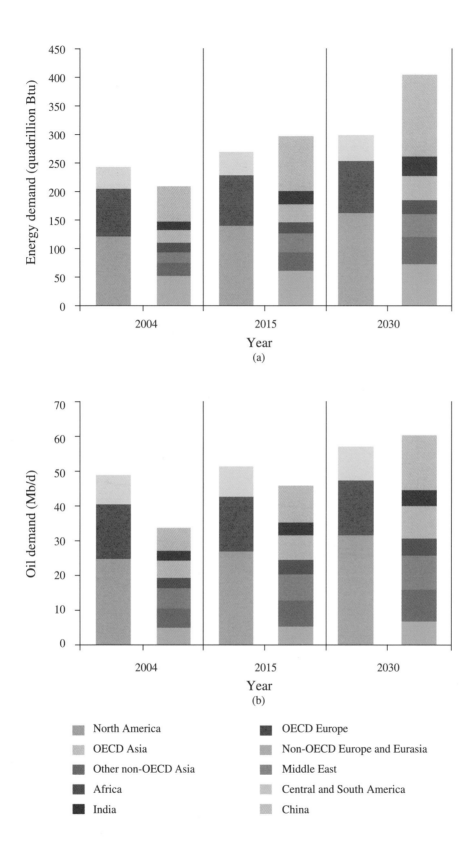

Fig.1-3 2030 Energy (a) and oil (b) demand trend of global economic entities[8]

Oil and Gas Resources in China: A Roadmap to 2050

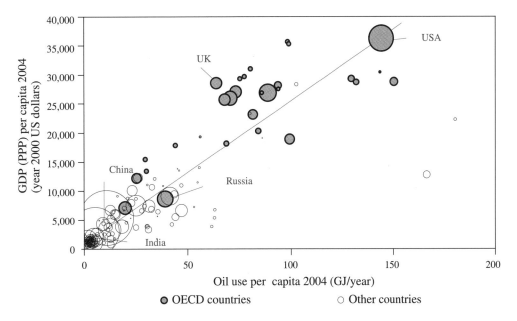

Fig. 1-4   Relationship between previous global oil consumption per capita and GDP growth[9]

With the trend of current economic globalization, it is necessary for our economic development to participate in the international competition, share the global oil and gas resources and compensate the domestic oil and gas shortage for the import; however, the oil and gas import strategy is also curbed by many aspects. After our entering into the world oil and gas market, the international crude oil price increases with doubling; the United States and Japan regard

China as the biggest energy rival; the western oil companies forbid China from overseas oil and gas exploration with the imported software and technologies, which results in our frustrating and disturbing investment projects of overseas oil and gas; moreover, as our over 80% imported crude oil are from the Middle East regions, the transportation must cross over the Straits of Malacca where is a military "throat". Therefore, our energy security has been tremendously threatened.

For a large country like China, the principle which is rooted on the domestic oil and gas resources must be adhered to guarantee the stability and even improvement of the oil and gas self-sufficiency degree with the economic growth, maintain our political and diplomatic independence of sovereignty, and guard the national benefit. Based on this principle, we should positively expand the exploration of the overseas oil and gas resources and properly import the oil and gas to protect our domestic oil and gas resources, keep the high-speed national economic growth, and maximize the national benefit. President Jintao Hu pointed out on the Central Economic Work Conference in the end of 2003 that "the international oil crisis seriously impacted on some important oil importer and exporter countries, so we must learn the lesson from their experiences and take the precautions". He indicated again on the National Science and Technology Conference in 2006 that "the innovation capacity must be improved" and "the energy and water resource development must be taken as the priority and the major bottleneck issues which restrain the economic development must be solved decidedly". With the instruction of Premier Jiabao Wen, "China Sustainable Development Report on Oil and Gas Resources Strategic Research" [3] finished by Chinese Academy of Engineering in 2004 concluded that "rooting in the domestic resources" and "the fundamental guarantee function of the domestic oil and gas resources" should be the primary principle which must be followed for the sustainable development of the oil and gas resources. The state council stressed again in the "Decision on Strengthening Geological Work" on 20th January 2006, that "the energy and mines are the significant strategic resources, which must be the priority of the geological explorations".

1.2 Oil and gas resources and potential

1.2.1 Total oil and gas resources and distribution

There are relatively abundant conventional and unconventional oil and gas resources on the planet. The assessment results of the United States Geological Survey [11] show that the recoverable resource volume of the global conventional oil is 413.8 billion t, the total volume of recovery is 134.4 billion t and the recovery percent is 29%; the recoverable resource volume of the natural gas is 436 trillion m^3, the total volume of recovery is 74 trillion m^3 and the recovery percent is 17%. The *World Energy Statistics Report 2009* newly issued by BP Company [12] presents that by the end of 2008, the proved reserves of the global remaining oil are 170.8 billion t, the output is 3.93 billion t and the reserve-production ratio is 42; the proved reserves of the global remaining natural gas are 185.02 trillion m^3, the output is 3.07 trillion m^3 and the reserve-production ratio is 60.4. In addition, *Oil & Gas Journal*[13] reports that the proved reserves of the global remaining oil in 2007 increased more 737 million t than the last year, even with the rising output, which was about 1.2% of the proved reserves of the global remaining oil and mainly from Venezuela, Brazil, Saudi Arabia, Kuwait, Iran, Angola and other regions. However, some experts remind that the human beings have consumed about 1/3 oil reserves so far, which were relatively light, easily mined and processed; while we are facing the current oil resources which mostly have the bigger specific gravity, higher acidity and hard-processed oil; therefore, more energy consumption and environmental issues would be produced in the oil processing.

Geological Resources and Recoverable Resources

The geological resources refer to the ultimately provable oil and gas total with the current available technologies, including the proved and yet unproved; the recoverable resources indicate the recoverable oil and gas total with the predictable technologies, covering the mined.

The distribution of the oil and gas resources in the world is extremely uneven. The remaining oil and gas resources are mainly located in Middle East regions, Community of Independent States (CIS) and North America. For instance, the oil reserve of Middle East regions in the end of 2007 was 102.5 billion t, the North America's was 28.9 billion t, the former Soviet Union and Eastern Europe's was 13.7 billion t, and the Central and South America's was 15.1 billion t[14]. In the unconventional oil and gas resources, the coalbed gas is 190 trillion m^3, which is mainly located in North America and Russia; the heavy fuel oil is 110 billion t, oil-sand oil is 131 billion t, and shale oil is 185 billion t, which are mainly located in America; the tight sand gas is 91 trillion m^3, and hydrate gas is 500 trillion m^3, which are mainly located in the North America and Russia.

For our resources, the new round of resource evaluation results [15] in 2005 showed that our oil and gas resources were also relatively abundant. The perspective reserves of oil and gas were respectively 108.6 billion t and 56 trillion m^3. Nowadays, the development of our offshore oil and gas exploration is limited in the coastal waters, covering 15.2 billion t oil reserves (14.0% of the national resource reserves) and 13 trillion m^3 natural gas reserves (23.2% of the national resource reserves); the current proved recoverable oil volume is 2.9 billion t and the recoverable natural gas volume is 5 trillion m^3, which are respectively 19.1% and 38.5% of the perspective resource reserves.

1.2.2 Oil and gas resources potential

As early as 1956, M. K. Hubbert, the famous United States petroleum geologist had pointed out the peak oil which indicated that the global oil output would reach the maximum at a certain time and then yearly reduce till zero supply. This is the "bell-shaped curve" [16].

Since the first industrial oil well successfully built in Pennsylvania, United States in 1859, the whole 150-year development of world modern oil and gas exploration has shown that the discovery and production of the oil seem to have no ending with the growth of the exploration volume and improvement of technology, but the exploration amount and output will gradually reduce. In the past 150 years, most of the oil reserves in the world were discovered in the 10 years from 1956 to 1965, including Daqing Oilfield in China; since then, the proved oil reserves started to drop, but the consumption increased yearly (Fig. 1-5). Since 1980, the annual consumption has been bigger than the proved recoverable reserves.

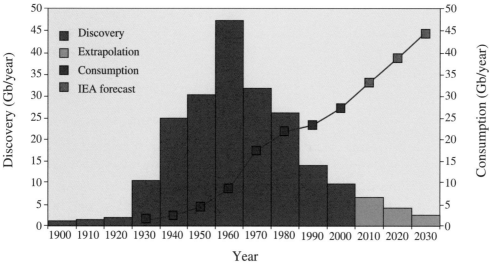

New discoveries from 1995 till 2025 is 100 billion barrels found and 100 billion barrels expected to be found,USGS mean prediction for the same time period is 649 billion barrels.

Fig. 1-5　Development and forecast of 1900–2030 world oil need and supply[6]

The relationships between the oil exploration and production-consumption of some oil-rich developed countries are always regarded as the models for the forecasts of the world oil production and consumption. Based on this, M. K. Hubbert had predicted that the oil output peak of the United States would appear from 1967 to 1971, and then would reduce gradually after this. The fact is that the United States exploration peak appeared in 1935 and the output peak (Peak Oil) would occur after 35 years, i.e. 1970[16] (Fig. 1-6). During this period, the United States experienced the Great Depression, World Oil Price Volatility and other major issues, which all obviously impacted on the oil and gas exploration and production. However, the statistics reflected that the oil well number in these 35 years was much more than the number in the past 100 years, but the oil and gas exploration and discovery volume of this period was gradually reducing.

Hubbert's successful prediction about the United States peak oil evoked the interests of the scientists and research agencies on the peak oil issue. C. J. Campbell and J. H. Laherrère[2] published article about the world peak oil forecast on *Scientific American* in 1998 when the world oil price was depressed, and proposed that the global cheap oil era would be ended. L. R. Brown[19] warned in 2003 that human beings should take the solution research as soon as possible as the world oil output would start to drop in the coming 10 years. Currently, Norway seemed to undergo the similar experience to the United States (Fig. 1-7), which the proved oil reserve reached the peak in around 1978 and then the oil output continued to increase, but the exploration kept dropping. All the competent authorities predict that the oil output in Norway would present the decreasing trend (Fig. 1-7).

Peak Oil

It was proposed by M. K. Hubbert, the petroleum geologist, in 1956. He predicted that the world oil production would follow the bell-shaped curve, i.e. the production would stably increase to the peak and then rapidly reduce.

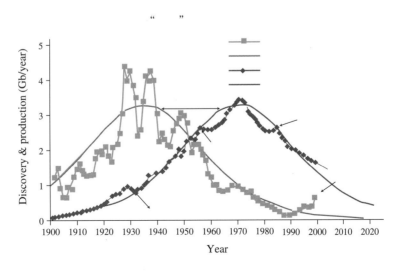

Fig.1-6   Delay period[6] of the United States oil exploration peak and production peak

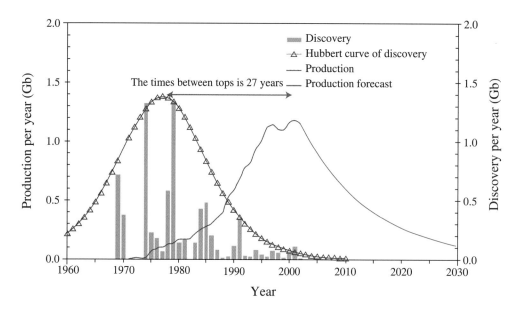

Fig. 1-7　Relationship forecast[6] of Norway oil exploration peak and production peak

Although the oil may exist in many regions where have not been drilled yet, the unexplored areas are rare and the exploration is very difficult. IEA [20] pointed out in the report on the world energy prospect in 2006 that "in view of the discovery and future production of the world oil, the natural gas liquid (NGL) production is expected to be increased; plus the subsea oil and polar oil, the global peak oil is possible to occur in around 2010 (Fig. 1-8), covering the United States, Europe, Russia and other regions except the Middle East regions where would keep the current output for a long term."

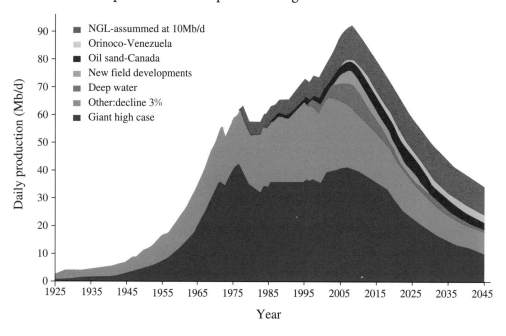

Fig. 1-8　Perspective world oil and gas resources production forecast (about 2050) [9]

Oil and Gas Resources in China: A Roadmap to 2050

From the above analysis, the global proved oil and gas reserves can basically meet the world's need till 2050. The Middle East regions will still be the main producer and supplier of the global oil and gas products. China will exceed Japan and become the biggest oil importer in Asia (Fig. 1-9).

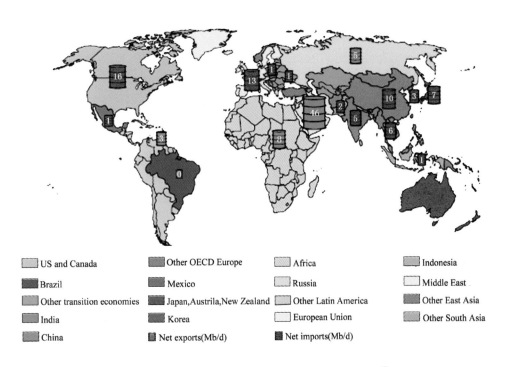

US and Canada   Other OECD Europe   Africa   Indonesia
Brazil   Mexico   Russia   Middle East
Other transition economies   Japan,Austrila,New Zealand   Other Latin America   Other East Asia
India   Korea   European Union   Other South Asia
China   Net exports(Mb/d)   Net imports(Mb/d)

Fig. 1-9   2030 world oil demand and supply forecast[9]

According to the forecasts of IEA[8] and EIA[10], Daqing Oilfield, the biggest oilfield in China, was discovered in 1959. In the following 25 years, other large oilfields were successively found, but after that no large oilfields were detected. If there were no new large oilfields, the oil output in China would reduce and the economy would keep rapid growing at the same time and the demand for the oil import would increase; however, in view of the growth and decline trends of the world oil and gas resources and demand, it may be impossible to provide the oil and gas for the world importer countries in the world after 2030.

Therefore, the development in China cannot continue using the economic growth models and energy consumption modes in the United States and other developed countries in Europe. The population of today's United States is only 5% of the world population, but consuming 25% of the world oil resource; the United States is trying to keep 25% of the perspective world oil growth over a long period, so it is estimated that it would be 959 thousand t/d in 2030. If China follows the United States current energy consumption mode and reaches its GDP per capita, the oil and other energy consumption would be an enormous figure. Even if all the productive oil and natural gas in the world are supplied to China, they are not enough!

On the other hand, our oil and gas reserves are relatively abundant and numerous unknown oil and gas resources to be discovered and verified. The report[21] of Sinopec (China Petrochemical Corporation)in 2003 showed that our oil and gas reserves of marine facies carbonate rock strata were over 30 billion toe and the discovered oil and gas reserves till 2005 was only 5% of the total reserves. There are rich oil and gas resources contained in the areas of new exploration or unexplored region, like marine facies carbonate rock strata, deep basins and deep waters, which reserves and potential need to be further studied and estimated. In addition, with the progress and the human improvement of science and technology, some potential oil and gas resources may be available, such gas hydrates, oil sand/asphalt mines, oil shales, shale gas etc.

Compared with the overseas advanced countries, the verification rate of our oil and gas resources is lower. The verification rate of the national average oil recoverable resource is only 37% and the natural gas is only 18%[15], which are much lower than the world averages[22] (73% and 60.5%). There are still 63% recoverable oil resources and 82% recoverable natural gas resources to be discovered[15]. With the instruction of the Continental Origin of Petroleum Theory, the resources in our eastern basins have been well developed and the proved oil and natural gas are 38.9% and 23% of the total reserves; the proved oil and gas resources in the western areas are only 18.9% of the total reserves due to the relatively low research level. The oil and gas exploration is very potential.

The recovery ratios of our various oilfields are commonly low. The average recovery ratio of our onshore oilfields is 30.1%[23], which is much lower than the average (50%) of the advanced countries; the recovery ratios of the offshore oilfields, low-permeability oilfields, fault-block oilfields and heavy oilfields are far lower. With the recent growing of the international oil price, the overseas big petroleum companies have confirmed the final oilfield recovery ratio to over 70%. This shows that our recovery ratio improvement is great potential.

Owing to our complex oil geological conditions and low oil and gas exploration level, the peak of our oil and gas production seems not coming and the peak intervals of the oil and gas exploration and production should be relatively long. Therefore, facing the urgent oil and gas demands of our economic development and comparing with the exploration effects of overseas high-exploration countries and regions, it is completely possible for us to develop the new oil and gas explorations, form the new oil and gas geological theories and establish the complete exploration technological framework, in order to improve the exploration level and effect and guarantee that the domestic oil and gas outputs can reach over half of our oil and gas consumption in the coming 20–50 years with the support of the whole-society comprehensive energy-saving measures.

> ### Recovery Ratio
>
> The recovery ratio of the crude oil refers to the percentage of the recovery of the crude oil in the geological reserves.

1.3 Development of our oil and gas resources exploration

Over the half century, with the instruction of the Continental Origin of Petroleum Theory, the development of our onshore basin oil and gas exploration has obtained great achievement and entered into the oil stable growth stage and natural gas rapid development stage. However, the overall oil and gas exploration is still lower than those overseas. The remaining oil and gas resources are abundant, which are mainly located in the rock strata, marine carbonate rocks, deep basins, foreland basins, and deep waters. The oil and gas exploration is facing a series of difficulties, such as the unclear new-type geological distribution of oil and gas reservoir, harder identification and forecast of the rock strata, deep structural traps, foreland basins, marine carbonate rocks and others, further supporting technologies enhancement of the complex surface topography and complex tectonic regions etc. The domestic crude oil output keeps stably growing and enters the "double-high" stage. The situation and challenges of the oil and gas development: it is difficult to improve the recovery ratios of the old oilfields whose moisture is over 85%; many low-permeability and ultra-low-permeability reserves are hard to be economically and effectively used; the economic recoveries of the middle and deep heavy oil are not easy; the new gasfields are more complex and the production safety has tough situation; however, the improvement of the success rate and recovery ratio of the oil and gas exploration are still potential with the advanced exploration technologies.

With the ever deepening of the oil and gas exploration and the diversity and complexity of the exploration objects, the requirements for the exploration target forecast accuracy and evaluation accordance rate are obviously rising. Our oil and gas exploration presents the development trends from the structural trap to rock strata trap, from simple structure to complex structure system, from shallow to deep and ultra-deep, from continental clastic rock-oriented exploration to marine carbonate strata and volcanic rock exploration and natural gas exploration and mature basin refining exploration, and from shallow waters to the deep waters and ultra-deep water.

Compared with the overseas countries, our onshore oil geological theories and exploration technologies are always leading and form the theories like "Source Control", "Multiple Oil and Gas Accumulation Zone"[24]. The reservoir forecast accuracy can reach 5–8m and the exploration target evaluation accordance rates for different basins can achieve over 60–80%. The systematic oil geological theories and supporting technologies and methods have been formed and the 3-level evaluation system (region, zone and target) has been established in the eastern Bohai Bay basin and Songliao basin; especially the recent lithologic stratigraphic reservoirs, zone-trap-accumulation controlled by the structure-sequence and other geological theories, which construct the 2 core technologies (continental sequence stratigraphy and high-resolution seismic), promote the rapid development of the continental strata reservoir lithology exploration, and provide the technological support for the discovery of the oil and gas verification reserves.

The establishment of the Continental Origin of Petroleum Theory and successful exploration of the continental basins have led our oil industry to the independent development. When the oil and gas reserves in the continental basins like eastern Songliao basin and Bohai Bay basin are hard to ensure the improvement of the oil and gas output, China properly proposes the strategy of "stabilize the east and develop the west" for our oil industry. The exploration focus starts to transfer from east to west (regional); and from continental strata to marine strata (bed series). The target of these transfers is the marine carbonate rock strata.

The marine carbonate rock strata are a significant exploration field in the world. The statistics of C & C in 2000 showed that the oil and gas reserves in the global marine carbonate rock strata could be 38% of the total reserves and the oil and gas reserves in the large oil and gas fields could be about 60%. Our marine carbonate rock strata have wide distribution, big remaining thickness and much oil and gas indicates. Since the 6th Five-Year Plan, we have carried out researches on the marine carbonate rock strata many times, but no major breakthrough can be achieved due to the objective conditions limits, such as the complexity and exploration difficulty of our ancient marine carbonate rocks, which led the large different awareness of our geologists. The very low level reality of geological research and exploration of the marine carbonate rock strata in our county greatly restricted the awareness of the potential oil and gas resources of carbonate rocks strata.

With the proposal of the "Second Venture of Oil and Gas Exploration", the research and practice of the oil geological theory of the marine carbonate rock strata have been deepened. We discovered Puguang Gasfield (450 billion m³ proved natural gas reserve) in eastern Sichuan in 2003; successively found a series marine carbonate rock strata gasfields in Sichuan basin, like Yuanba and Longgang; found Tahe-Lunnan carbonate reservoir oilfield in Tabei uplift with proved, controlled and predicted 1.5 billion t reserves; and discovered the carbonate rock oilfields and gasfields in Tazhong uplift and Bachu uplift. These exploration results show that our pre-cenozoic marine carbonate rock strata have wide exploration perspective and huge potential with the progress of the theories and technologies and are the significant part of our oil and gas resources strategy.

Second Venture of Oil and Gas Exploration

Facing the severe oil security situation, Guangding Liu, the member of Chinese Academy of Sciences, submitted the "Proposals on the Second Round of Oil and Gas Exploration in China" to the State Council in August 17th 2001 and presented the constructive opinions on the second exploration of Chinese oil and gas resources. On August 27th, Jiabao Wen, the vice Premier at that time, instructed that "the strategic exploration of the oil and gas resources should be emphasized and the new breakthrough in the pre-cenozoic marine carbonate rock strata should be achieved". Therefore, the new era of the Second Venture of Oil and Gas Exploration in China has started.

Since the proposal of the "Second Venture Exploration", we have obtained many achievements in the remaining marine basins, for example, ① high- yield crude oil in Ordovician layer of Shengli Oilfield Shenghai G2 well with daily output of 1,059t and a series of ancient buried hills in Chezhen and drilled high-yield crude oil; ② Tahe-Lunnan Oilfield, the biggest carbonate rock oilfield discovered in Tarim basin, which the major reservoir is the Ordovician carbonate rocks and has high non-uniformity, 4,100–4,800m deep buried oil reserves, 632 million t proved oil reserve, 72.55 billion m^3 proved natural gas reserve, and 1.46 billion t 3-grade oil and gas equivalent; ③ the large Feixianguan Formation Gasfield in northeastern area of Sichuan basin, whose reservoir layer is Triassic Feixianguan Formation oolitic dolomite, with hundred billions cubic metres of proved gas reserves in many large gasfields, like Puguang and Luojiachai; ④ the new fields in the lower part of combination in superimposed basins, which are the recent important discoveries in China, like the deep layer Songliao basin, buried hills in Bohai Bay basin, Weiyuan Cambrian system in Sichuan basin and Ordos basin; ⑤ a series of significant discoveries, like Kela 2 Gasfield and Dina 2 Gasfield, in foreland basin thrust belt exploration, which are mainly the anticline oil and gas reservoir with high abnormal pressure, large scale, high abundance and high output.

We have southeast coastal sides with wide water area. In the recent 25 years, our ocean exploration has oriented to the coastal waters. We have mature knowledge and exploration engineering technologies for the landlocked sea and shelf extensional basin oil geology, effective oil and gas exploration, rapid-growing oil and gas outputs, and relatively world-level theory and exploration technology for the shallow sea shelf extensional basin oil and gas geology. The recoverable sea oil and natural gas reserves are respectively 7.21 billion t and 10.7 trillion m^3; the proved recoverable reserves are 490 million t and 0.26 trillion m^3, the remaining recoverable reserves are 6.72 billion t and 10.44 trillion m^3. In 2006, the annual outputs of the oil and gas were respectively 27.83 million t and 6.9 billion m^3; 60% of the proved sea reserve was heavy oil and the water displacement recovery was only 18–22%.

However, our sea oil and gas discoveries are mainly the structural-type oil and gas reservoir of the offshore basins, the deep-water and deep exploration and lithologic deposits exploration are just beginning, and the understanding of the oil and gas reserves and exploration potential is not clear. Our exploration of the continental slope under 300m-deep water just starts and there is no exploration of the deep water basin. The restraints are the low fundamental geological research level, like the structure characteristics, sequence stratigraphy development characteristics of the deep-water basins and no major equipments and construction technology for the deep-water oil and gas field's exploration. The knowledge and theory for the deep-water basin oil and gas reservoir under our unique geological conditions have to be set up. The development direction of our ocean oil and gas exploration should be the complex oil and gas reservoirs like deep-water oil and gas reservoirs, lithologic oil and gas reservoirs and high-temperature and high-pressure oil and gas reservoirs. With the theoretical innovation and technological breakthrough, the ocean oil will be one of our oil and gas reservoirs and growth points.

The oil exploration is regarded as one of the most difficult engineering projects, while the difficulty and complexity of the oil exploration in China are rare in the world. With the development of the oil and gas exploration, the target has been transferred from the simple structural oil and gas reservoirs to the complex cryptic reservoirs with deep burial depth, complex geological conditions and strong non-uniformity reservoirs. The current theoretical models are relatively simple and strongly restrained and the seismic reflection which is as the major technology for the oil and gas exploration is hard to form the image of the complex oil and gas reservoirs; therefore, the exploration cost and the risk are high, the success rate of the test well is low and the growth of the oil and gas reserves is slow.

Currently, the key to the oil and gas exploration technology breakthrough depends on whether the geophysical theory and method can achieve huge progress. Our continental oil and gas basins have many characteristics, such as the multi-material sources, multi-deposition systems, big aqua dynamics changes and high-frequency water advancing and retreating. So the clastic reservoirs have the following basic characteristics: polycyclic and multi-layer reservoirs formed among the thin sand shales. The characteristics of our offshore oil and gas basins are deep burial, big impact of the later tectonic movements, reefs, caves and fractured reservoirs. The data analysis shows that the contribution rate of the geophysical prospecting for the oil and gas resources/finished oil conversion is about 50% which is the highest. Therefore, the geophysical prospecting, especially the reflection seismic exploration techniques, are the real bottleneck for the oil and gas discovery. With the instruction of the regional geological theory, the vital scientific issues which can settle down the complex oil and gas exploration will be to fully understand the propagation characteristics of seismic wave fields, seek the fundamental theoretical research on the complex oil and gas reservoirs' geophysical data collection, analysis and processing. Meanwhile, the emphasis of the geophysical techniques such as gravity, magnetism and electricity and the deep research of the comprehensive geophysical methods, theories and technologies for the complex geological entities are also the important paths to solve the oil and gas exploration technological issues. Therefore, the oil and gas geophysical fundamental theories and key technological methods study are the urgent requests of the oil industry and geophysical exploration development.

The beginning of the oil seismic exploration techniques development is the improvement of the signal to noise ratios of the seismic prospecting, from the spot records, analogue magnetic tape recorders to the digital tape recording, from the single profile to the multiple superimposed tectonics, from the 2-dimensional seismic to 3-dimensional seismic, the seismic instrumentation from the scores of channels to the current thousands of channels. However, the lithologic stratigraphic reservoirs require to enhance the resolving power, besides the improving the mater frequency, widen the bandwidth and transforming from the 16-bit sampling to 24-bit sampling, but the current technologies still cannot solve the more and more complex geological exploration issues. Therefore, the next development direction should still focus on improving the signal to noise ratio and resolving power of the seismic exploration technology. However, the self defects of the seismic data acquisition system has restrained the seismic technology development and become the bottleneck of the oil and gas exploration technology development.

The current oil and gas exploration must go to the deeper and more complex geological structure and lithologic strata and obtain the weak effective information under the strong interference, so the high-accuracy and high-solving resolution data are possible to deepen the understanding to the oil and gas reservoirs, which also propose the new requirements for the seismic recording system study and innovation. The existing seismic data acquisition system has 2 major issues: the geophone (the dynamic sensor will transform from the mechanical vibration to electrical signal and the dynamic range is only 68dB) cannot match with the instruments, which appears the small instantaneous dynamic, low sensitivity and narrow bandwidth etc.; the timely receiving channel number is too less, which cannot meet the dense-point collect needs.

The domestic and overseas oil exploration experiences also show that the applicability of the existing geophysical exploration techniques has strong geological features, like the geophysical techniques for Daqing Oilfield cannot discover Shengli Oilfield; the techniques for Shengli Oilfield cannot find Kela 2 Gasfield; the migration and imaging techniques which are suitable for the salt dome exploration in Mexico Gulf cannot apply for our offshore carbonate vug oil and gas reservoirs exploration. With the more and more complex exploration conditions, the existing geophysical exploration theory is facing serious challenges.

"The basic of the geophysics is the observation and the geophysical instruments are the key to the observation." The new-generation digital geophone and large-scale high-accuracy seismic acquisition system should be developed to meet the complex geological oil and gas exploration and realize the breakthrough of the new seismic technologies.

AAPG Annual Conference, the world-famous conference on oil and gas exploration, proposed for the oil and gas exploration situation and development trend in 2006: ① successful business strategies; ② learning from exploration and exploitation successes, failures and mistakes; ③ perfecting the search for unconventional plays and technology (including coalbed gas, cracks clastic, carbonate rocks, unconventional reservoir, non-seismic methods, ultra-shallow natural gas reservoirs etc.); ④ characteristics of the world giant oil and gas fields; ⑤ studies on the integration of the geology and geophysics and engineering methods[26]. In 2008, the conference added the studies on the deep-water slope basin system, reservoir description and modeling, oil and gas system and basin analysis, oil-hydrocarbon and coal-formed hydrocarbon etc., especially strengthened the investment for the environmental issues related with the resource development[27]. Our situation has something common with the topics and difficulties of the world oil and gas exploration and also our own unique issues, so we must aim at our real geological situation and characteristics and form the self-innovative theoretical system and core technologies to provide solid technological and theoretical supports for the stable development of our oil industry.

1.4 Influence and position of oil and gas science and technology development in oil and gas resources

The recollection shows that every major breakthrough of the oil industry has been closely connected with the important development of the geological theories, technologies and equipments; and its development and evaluation are rooted from the scientific and technological progress. The over 100-year world oil industry development is an innovation history of the oil science from the primary wildcat drilling and blind extraction to the modern digital oilfield. The world oil industry has realized a qualitative leap.

Digital Oilfield

The digital oilfield should take the oilfields as the research objective, the computer and high-speed internet as the carriers, the spatial coordinate information as the reference; integrate the data of the oilfield production and management; adopt the technologies like simulation and virtuality to visibly express the multi-dimensional data based on the optimization modes built for the oilfield production and management processes; realize the horizontal coverage of the whole oilfield and vertical multi-level information positioning from the oilfield surface to the underground; improve the overall information analytical capacity; support the vital business like the oilfield exploration and development; overall supplement the decision-making analysis of the oil operation management; further cultivate the potential values of all the links; and create a solid information-support environment for the sustainable development of the oilfield enterprises.

1.4.1 History of World Oil and Gas Sciences and Technologies

The development of the world oil and gas sciences and technologies express from the oil and gas geological theories, oil and gas exploration technologies, oil and gas development technologies and oil and gas drilling processing and other aspects. The division of the world oil and gas science and technology development stages in this book is based on the development of the oil and gas geological theories and exploration technologies and other aspects. Since the first oilwell[17] with modern industrial significance drilled by Drake who was an American in Pennsylvania in 1859, the development of the world oil and gas science and technology has experienced 4 stages and 3 great leap-forward growths, which has brought 3 leap-forward growths of the world oil reserve and output.

First stage: oil and gas seepage to find oil and gas (1859–1890)

The early-stage oil and gas exploration was based on the oil and gas seepage. This period was from 1859 to the early of 1890s when the anticline theory was formally found. It was the foundation stage of the oil and gas exploration theories, as the theories had not been set up and the major content of the oil and gas exploration was to investigate the oil and gas seepage and drilling.

The United States and Russia were the big oil producers at that time and the earliest countries which found oil and gas fields by the seepage. With the oil and gas seepage, the United States successively found the oilfields in Appalachians, Cincinnati, Rocky Mountains, California and Central Platform and other regions. At the end of 19th century, the overall oil output in the United States reached 8.6 million t. Russia drilled the first oilwell in Aqua, Ciscaucasia in 1864 and found Kudak Oilfield and then the Balahane Oilfield and Bibi Ebbutt in Baku etc. At the end of 19th century, the annual oil output in Russia reached 10 million t and was the first oil producer in the world at that time. In addition, other countries like Mexico, Venezuela and Peru also carried out survey based on the oil and gas seepage and found oilfields[28] during the late 19th century to the early 20th century.

Oil and Gas Seepage

The oil and gas seepage is referred to the underground gathered oil and gas escaped for certain reasons and floated to the surface. It is the mark of the oil and gas generation and gathering. The normal oil and gas seepage is earth wax, asphalt, oil spring, salse, oil sand etc.

Although the oil and gas exploration in this period lacked of the theoretical support of the oil and gas geology, some primary knowledge had been formed, like the awareness of the relationship between the oil and gas reservoir distribution and anticline, oil source from the marine strata and etc. It was a very important era in the world oil industry history and accumulated precious experience and raw data for the following oil and gas geology theory.

Second stage: anticline to find oil and gas (1891–1953)

In the previous oil and gas seepage survey, people had vaguely found the certain relationship between the oil and gas seepage distribution and geological anticline. With the expansion of the exploration areas, the awareness became clearer. In 1861, E. B. Andrews[29] and R. S. Hunt[①] obtained the conclusion of oil distribution along the anticline based on their geological research and then further developed it to the anticline oil statement. In 1885, I. C. White and others from the United States Geological Survey, summarized the oil organic production, impacts of the porosity and permeability of the reservoir bed on the reservoir properties and anticline oil etc., which provided basis [30] for the anticline statement.

---

① Hunt T S. Bitnmens and mineral oils. Montreal Gazette,1861.

Anticline

The anticline is the structure formed by the fold deformation of the sedimentary strata. A complete anticline structure shows the round to oval and the side strata bent downward.

In 1891, based on the anticline statement, the United States found a series oilfields in Kern River, Midway Sunset and Micktric during the oil and gas exploration of the San Joaquin Basin, California, which symbolize the non-theory support for the oil and gas seeking era was over and human beings entered the anticline statement stage. This period had lasted over 60 years until the middle of 1950s. The huge impact of the anticline statement was due to the long-term high success rate and easy identification and exploration. Although the anticline statement is a past, being an important oil and gas trap type, the anticline structures are still the first targets in some certain areas (like compression basins) or low exploration rate regions.

The establishment and successful instruction of the anticline statement have become the first leap of the world oil industry history and shown the huge impact on the global oil and gas exploration. The contribution of the anticline statement to the oil geological theory was not only important oil and gas trap type, but also the oil and gas accumulating concept, which facilitated the birth of the oil geology and helped it separate from the traditional geology. In 1915, I. C. White proposed the fixed-carbon ratio theory which believed that the oil and gas perspective of a region could be judged by the metamorphic rocks tested by the fixed-carbon ratio [17]. In 1915, D. Hager [31] published *Practical Oil Geology*, the first oil geology academic work in the world. In 1917, the American Association of Petroleum Geologists (AAPG) was established, which largely accelerated the oil geology generation. Mobile Oil of New Jersey, the predecessor of Exxon, the biggest oil company in the world, was also established in 1917. Thereafter, some classic works of oil geology were successively published, which symbolized that the oil geological theory was basically formed and the oil geology was born. Meanwhile, the petroleum geochemistry and micro paleontology were also extensively adopted by the oil geologists. In 1950, the oil and gas migration, gathering physical simulation and mathematical simulation methods were applied for the petroleum geology theoretical studies [28].

When the anticline theory was used in the oil exploration, the world oil and gas exploration technologies also developed rapidly.

Exploration: After the heavy, magnetic technology in the oil and gas exploration, the seismic exploration techniques aimed to the anticline structures in the coverage were also developed. In 1924, the US first adopted the refraction method to find the Salt Dome Oilfield [28] in Mexico Gulf, and first used reflection method to find Maud Anticline Oilfield [32] in Semino Basin, Oklahoma in 1928. The success was regarded as the first revolution of the seismic exploration technology. Thereafter, the resistivity well logging technology and rock coring technology were also applied to the oil exploration.

Drilling: As the rotary drilling rig replaced the cable rig, the drilling depth in 1931 was to 3,000m deep and 5,000m deep in 1945. In 1949, the first 6,000m-deep test well in the world was drilled in Wyoming, United States.

Exploration territory: Besides the continental oil and gas exploration, the human beings started the shallow-water oil and gas exploration.

With the establishment of the oil geology theory and development of some exploration technologies, the world oil and gas industry obtained rapid growing. In 1943, the United States annual oil output was 200 million t and natural gas 100 billion m^3. By the end of 1950s, a lot of oil and gas fields were found in the world and the annual oil output exceeded 500 million t.

Third stage: trap to find the oil and gas (1954–1990)

With the ever improving of the oil and gas exploration territory development and exploration degree, the success rate of the anticline exploration was reducing, especially in some stable-structure regions, which resulted in the doubt of the some oil workers to the anticline statement. In fact, as early as the beginning of 1920s to 1930s, there were already many non-anticline oil and gas fields discovered in the United States, such as the Hugoton great Gasfield found in Anadarko Basin in 1922 and Great East Texas Oilfield found in Mexico Gulf Basin in 1930, which were all controlled by the stratigraphic traps. Based on the doubt of the anticline statement and the summary of the new exploration experience, A. I. Levorsen proposed concept of "Stratigraphic Trap" in 1936, which covered stratigraphic trap and lithologic trap. But it did not interest people at that time. Until 1954, A. I. Levorsen[33] published *Geology of Petroleum* based on the further development of the stratigraphic trap and pointed out that "any rocks which can store oil, no matter the shape and formation, can be regarded as trap. The basic characteristic should gather and store the oil and natural gas". The publication of this book was immediately supported by most of the oil geologists and became the symbol of the trap statement replacing the anticline statement. This was the second leap of the world oil industry history. On the 20th International Geological Congress (IGC) held in 1956, the oil geology was listed as one of the 3 themes[34].

Trap

The places where are suitable to assemble the oil and gas in the underground strata of the oil and gas contained basins can be called as traps. There are 3 factors: ① the reservoir with the fluid accommodation space and percolation; ② caprock for the oil and gas escape upward;③ cover condition for oil and gas migration. The cover condition can be the deformation of the caprocks and the traps formed by the fault, lithology changes and others.

After the Second World War, the oil and gas became very important for a nation's economy and military. The progress and development, such as the computer technology, remote sensing techniques, deep-sea drilling technology, seismic technique, plate tectonics, kerogen hydrocarbon theory etc., largely facilitated the world oil and gas exploration growth.

The oil and gas exploration hotspots turned from the Western Hemisphere to the Eastern Hemisphere, from the continent to the sea, and from the domestic to the overseas. As the higher exploration degree of the domestic old oil and gas areas and lower discovery rate of the new oil and gas fields, the big oil companies of the western countries started to seek the chance overseas, especially the Eastern Hemisphere where had relatively low exploration degree. The new exploration hotspots were firstly the Persian Gulf, then Indonesia of South-Eastern Asia, Libya of North Africa, Nigeria of West Africa. The reserves and output of these regions rapidly increased due to the discovery of many new oil and gas fields. In addition, the Former Soviet Union found a series large and giant oilfields and gasfields, such as Orenburg, Urengoy, Tengiz in Volga-Ural, West Siberia, East Siberia regions and Kazakhstan. During this period, the climax also occurred in offshore exploration, besides the large oil and gas fields found in the Persian Gulf, the United Kingdom, Norway, Denmark, Sweden and other countries also joined the North Sea oil and gas exploration team and found a lot of large oil and gas fields like Ekofisk. In 1968, the United States found Prudhoe Bay Oilfield in the north slope of Alaska within the Arctic Circle. Since then, except the South Pole, all the continents, polar region, deserts and seas have been the places where people seek oil and gas. Until 1990, the world remaining oil reserve was 137.2 billion t and the oil output was 3.02 billion t, which were respectively the 2.6 times of the world remaining oil proved reserve (53.7 billion t) and 1.9 times of oil output (1.63 billion t).

With the flourishing of the world oil exploration practice, the oil geology theories also entered the rapid development period. Based on the oil geology system with center of the oil and gas accumulation built in 1950s and combination of the related branching subjects of the geologic, many professional disciplines were born, such as the petroleum geophysics, organic geochemistry, reservoir sedimentation, seismic stratigraphy, lithofacies palaeogeography and tectonics, which closely related with the oil and gas exploration[28].

During this period, the rising of the plate tectonics had great impact on the oil geologic development. The theory was originated from the continental drift hypothesis [35], developed with the seafloor spreading [36], and completed as a theoretical system in the end of 1960s, which brought the activity theory into the earth science research [37]. The plate tectonics systematically describes the formation of the oil and gas contained structure and relationships among the oil and gas generation, migration and gathering, states the global oil and gas distribution rules, and powerfully instructs the oil and gas exploration. The biggest influence on the oil geological research is the oil and gas contained basins studies were transformed from the geological description to formation explanation, the basin classification and corresponding basin modes were also formed.

The confirmation of the late organic matter oil theory with center of the kerogen degradation of hydrocarbon was the major scientific progress which fundamentally impacted on the oil geologic in the late 1960s [38]. Through the systematical analysis, the scientists confirmed that the oil was the outcome of the thermal degradation of kerogen and formed with a certain temperature scope ("liquid hydrocarbon window"). The theory presented the mechanism and process of the organic matter of the deposit evaluation in view of the chemistry dynamics. With the instruction of the theory, a set of parameters which reflected the organic matter maturity in the aspects of geochemistry and optical measurement, such as the hydrogen index (I_H) , hydrogen-carbon ratio (H/C) , vitrinite reflectance (R^o) , time temperature index (TTI) , heat alteration index (TAI) , instant obtainable kerogen type, maturity, hydrocarbon potential and etc. The hydrocarbon generation theory and technological innovation provided the basis and methods for the hydrocarbon source rock study and hydrocarbon amount calculation. People started to root on the oil and gas generation and systematically considered the process of the oil and gas generation, discharge, migration and gathering, which greatly improved the success rate of the oil and gas exploration.

Kerogen

Kerogen was primarily used to describe the organic matter of the Scotland oil shale, which could produce wax-like viscous oil after distillation. Then it was used to refer to the solid dispersion of organic matter which was amorphous, acid-insoluble, alkali-insoluble and nonpolar in the sedimentary rocks. It is the outcome of the sedimentary organic matter with biodegradation and the parent material of the oil and gas generation.

In the early oil and gas exploration, as most of the oil was discovered in the sea strata, the oil geologists thought that only marine sediment would generate oil and gas. Zhongxiang Pan and our other senior geologists proposed the theory of continental oil generation in 1941 with the detailed studies on the oil and gas geology of Shaanxi, Sichuan and other regions, which extremely riched our oil theories. In addition, our oil geologists presented the "oil generation system" concept based on the continental oil and gas distribution characteristics, which was an entity composed of the oil resource, reservoir bed, caprock, trap and other accumulation factors which were connected with the oil and gas migration and gathering. Moreover, our oil geologists further showed the "source control theory" which referred to that the oil and gas accumulation was controlled by the hydrocarbon depression or hydrocarbon center and presented the zonal distribution. The oil system has its own characteristics and relative independence and forms one or many different-type oil and gas gathering zones, which can be overlaid and reconstructed each other and form the multiple oil and gas accumulation zone with continental basin characteristics.

With the promotion of the countries' emphasis and the breakthrough of the oil geological theories, all the oil and gas enterprises and relative agencies strengthened the research and development of the oil and gas exploration, which facilitated the rapid growth of this field and obtained many vital technological breakthroughs. (Fig. 1-10)

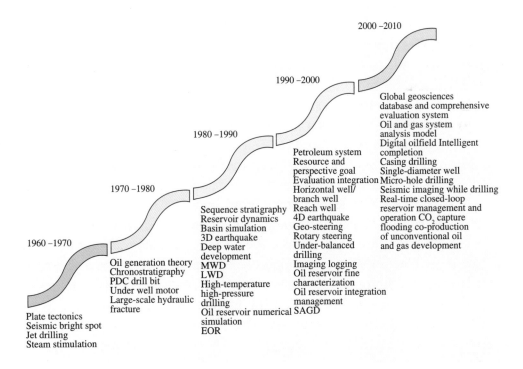

Fig. 1-10   History of part key technologies in world oil and gas industry [39]

Exploration: The transformation of the exploration concept also brought the huge development of the seismic exploration technology. In the 1960s, the digital recording replaced the traditional tape recording, which was regarded as the second revolution of the seismic exploration techniques, including a series of new technologies, like the computer for seismic data processing, migration technology for construction imaging, 1970s preserved amplitude processing for the oil and gas inspection, 1980s AVO technology for the reservoir pore fluid properties, working station replaced the manual interpretation etc.

The accuracy and detecting capacity of the seismic data were obviously improved; and the seismic technology could seek the structure trap, and discover the stratigraphic tap, lithologic trap and other subtle traps. The 3-dimensions seismic technology provided best observation method for the underground complex geological structure. Meanwhile, the well logging technology also rapidly developed, which finished 3 leaps in over 20 years: 1960s the digital logging replaced the analog logging; 1970s the computerized well logging replaced the digital logging; 1980s the imaging logging appeared. It should be noted that the synthetic seismic acoustic logging technology matured in 1979 solved the horizon correlation issue of the seismic reflection sequence and well logging data, created the comprehensive explanation for the geology, seismic and logging, integrated the seismic horizontal structure and the vertical high-resolution of the logging, adopted the inverse technique for the lateral projections of the target interest bed lithology, lithofacies and reservoir properties and placed the basis for the reservoir description[28,39,40].

Drilling: the jet drilling, directional drilling, optimization of parameters of drilling technology, PDC drill and foam cement cementing technology.

Developing: improving the oil strata pressure by pouring water, hydraulic fracturing technology, extraction steam stimulation for the remaining oil and heavy oil.

Fourth stage: petroleum system to find oil and gas (1991–now)

In the 1970s, some western scholars successively proposed "petroleum system" or similar concepts. From the end of 1980s to the beginning of 1990s, L. B. Magoon and W. G. Dow developed the "petroleum system" concept and contributed a lot to the practical utilization and standardization. In 1994, AAPG formally issued "The Petroleum System: From Source to Trap [41]", which was the third leap of the world oil and gas science and the symbol of this era.

L. B. Magoon and others divided the oil and gas exploration into 4 layers, i.e. basin, petroleum system, oil and gas zone and prospective trap. The basin study stressed the stratigraphic sequence and tectonic style of the sedimentary rock; the petroleum system studied the mature hydrocarbon source rocks distribution and the relationship with the oil and gas accumulation, which stressed the similarity of the geological traits of the zone-distributed traps; the prospective trap study stressed the current individual trap.

The petroleum system is a research layer between the basin and zone. If the scale is bigger, the quantity of oil and gas generated and accumulated is bigger and rate of oil and gas in the relative zone and trap is higher. The petroleum system is composed of the basic geological factors and accumulation roles which are necessary for the mature hydrocarbon source rocks and oil and gas accumulation. The basic factors include the hydrocarbon source rock, reservoir bed, caprock and overburden; the accumulation roles cover the formation of the trap and the generation, migration and accumulation of the oil and gas [42]. Only can the basic geological factors and accumulation roles match each other, the generated oil and gas can be accumulated.

The characteristics of the petroleum system show in 3 aspects: ① Geographical distribution characteristics, which are confirmed by the mature hydrocarbon source rock distribution scope, including the detected oil and gas fields, oil and gas show and oil and gas seepage; ②Stratigraphy characteristics, including the geological age distribution of the basic geological factors within the region;③Time characteristics, including the generation-migration-accumulation time of most of the oil and gas with the system (also called "vital time"). The traps of the pre-vital time or same-vital time are available and the post-vital time traps are generally unavailable, which can form the secondary reservoir under certain conditions. Currently, the petroleum system is studied and described with the burial history diagram, floor plan, sectional drawing, event graph and oil and gas accumulation table [24] related with the hydrocarbon source rock.

The sequence stratigraphy method set up by the late 1980s places the basis for application of the petroleum systems in the actual basins[43]. It adopts the sedimentation cycle the strata comparison units, the erosion surface and other bed boundaries which can show the sedimental hiatus events as the marks and forms the age stratigraphic framework to confirm the relations within the inner strata of the sedimentary sequence and the lithofacies distribution mode. With the analysis of the seismic, logging and outcrop data, it studies the relative sea level changes and division among it and the stratigraphic sequence and different sub-units within the sequence and distribution rules, which are controlled by the tectonic movement, eustasy, sediment supply, climate and other factors; researches their cause and effect relationships, interface features and facies distribution.

Since the 1990s, the world oil annual output has kept over 3 billion t. The output in 2008 was 3.648 billion t[12], but the remaining proved oil recoverable reserve can still keep slow growth(183.864 billion t in 2008, 0.8% higher than 2007) if there was no major discovery. The growth factors covered the improvement of the recovery ratio, expansion of the old oilfield and adding of new strata. Moreover, the update of the exploration concepts and the progress of the exploration technologies, which found a lot of small-size and hidden oil and gas fields, were also the reasons.

This period is the hi-tech era with the rapid development of the information technology, material technology, biotechnology etc. The important impacts on the oil and gas exploration technology are as follows:

Exploration: The 3-dimensions seismic technology became more and more mature in 1990s. As the development and application of the 3-dimensions processing, 3-dimensions pre-stack depth migration, 3-dimensions inversion, 3-dimensions attribute analysis, 3-dimensions coherent body, the quantity and quality of the structure and strata information offered by the seismic technologies reached an unprecedented height. The skillfully utilization of the imaging logging brought the interpretation and assessment level of the complex structure reservoir of the thin layer, thin interbed and complex lithology and crack to a new height[45]. The MWD could replace the conventional cable logging used in high risk, big gradient and horizontal wells [46]. The information offered by the NMR logging technology largely increased the strata assessment capacity of the well logging [40].

Drilling: With the computer technology and intelligent technology in the oil industry, the "revolutionary" update and breakthrough continually occurred in the traditional drilling processing; the processing technologies like the horizontal well, reach well, multi-branch well, underbalanced drilling, became mature; the technologies of the deep well, ultra-deep well and complex geological environment became perfect; the various new drilling liquids were adopted [39]. In addition, the series key drilling technologies like the expansion pipe, casing drilling, steel grain drilling, ultra-high-speed small motor, micro-foam drilling fluid and formative drilling fluids, were already used. Recently, the overseas countries are developing the laser drilling, micro drilling, continuous high-pressure jet drilling, new shock drilling and other advanced technologies.

Developing: layer explosion technology[47], polymer flooding technology, microbial enhanced oil recovery technology [48], and CO_2 flooding technology.

1.4.2 History of our oil and gas science development

When all the countries are seeking sea oil and gas, Zhongxiang Pan and other senior scientists proposed the "Continental Origin of Petroleum Theory" with Chinese characteristics according to our own actual geological situation. In 1941, Zhongxiang Pan [49] published "Nonmarine origin of petroleum in north Shensi, and the Cretaceous of Szechuan, China" on AAPG journal and pointed out for the first time that our continental strata could generate oil and form the oil reservoir. This was regarded as the start of the China "Continental Origin of Petroleum Theory".

With the instruction of this theory, a lot of giant oil and gas fields like Daqing, Shengli, Liaohe, Dagang and Huabei, were successively discovered in 1960s and 1970s, which became our significant oil production bases, placed solid foundation for our oil industry growth, and made indelible contribution as the "First venture of the oil and gas industry in New China" [25, 50].

Over the half century, the plate tectonics theory, basin analysis, sequence stratigraphy, reservoir dynamics, hydrocarbon generation theory, oil and gas system theory and etc. placed solid basis for the Cenozoic continental oil and gas resources assessment and oil and gas potential forecast; the 3-dimensions high-resolution, multi-wave seismic exploration technology, pre-stack depth migration, basin simulation, imaging logging, NMR logging technology and reservoir description technology provided effective ethnological support for the success rate improvement and complex and subtle reservoirs identification.

In 1980s, with the transformation from the structural hydrocarbon reservoir to lithologic oil and gas reservoirs and subtle reservoirs, the large-scale non-homogeneous sandstone oilfield exploration theory and sequence stratigraphy technology and others developed rapidly; the slice mining, water-control and stable-oil and tertiary recovery and other technologies were extensively applied, which further promoted our oilfield exploration level. The annual output of Daqing Oilfield kept 27-year 50 million t and realized high and stable production. The development of the seismic exploration techniques like the pre-stack depth migration recovered some old oilfield production and discovered new oilfields.

In 1990s, as the progress of the geological awareness and exploration technology, many basins in west were settled down. The western oil and gas equivalent output increased from 16.23 million t in 1990 to 72.28 million t in 2005.

In the end of 1990s, with some eastern Cenozoic oil and gas basins exploration entering into the middle and late stages, the pre-Cenozoic marine basin oil and gas resources buried under the Cenozoic basins gradually draw people's attention. Owing to their depth, complex structure and long evaluation history, the exploration theories and technologies were different from the Cenozoic continental sedimentary basins and the traditional theories and methods were restrained, but the potential was huge. Therefore, Liu Guangding pointed out that "the remaining pre-Cenozoic marine oil and gas resources development would be our main field in 21st century oil and gas exploration and the main characteristics of the second venture in the oil and gas industry" [25]. The comprehensive geophysical exploration techniques for the marine exploration, including the seismic imaging technology under complex surface conditions, residual basin of non-earthquake detection technology, igneous rock fracture extraction and etc. obtained long-term progress. The discovery of the marine oil and gas fields like Puguang Gasfield proved the significance of the exploration geological theory and technological progress for the marine oil and gas exploration.

1.4.3 Oil and gas science and technology development in geological theories, technologies and equipments

From the recollection of the abovementioned world and our oil and gas science history, the development of the oil and gas science and technology is mainly showed in the oil geological theories, exploration technologies and relative equipments.

The seismic exploration technology, for example, the conventional seismic exploration technology is the CMP stacking technique aimed at the horizontal layered media. For the complex structure, it commonly has the low signal to noise ratio and precision imaging and other problems. For solving the seismic imaging issued due to the lateral variation, the prestack migration imaging technique has been developed, which can overcome the defects of the traditional seismic treatment, improve the signal to noise ratio, have much better steep dip imaging than the traditional one, and provide the technological guarantee for the pre-stack attribute description reservoir like AVA (Amplitude Versus Angle)、AVO (Amplitude Versus Offset). In the real producing, Kirchhoff integral pre-stack depth migration has been extensively adopted. Furthermore, the pre-stack depth migration based on the wave equation has been developed, which has stronger keeping dynamics information capacity, calibrate the amplitude changes due to the focus and caustics and get verified in the practice. However, the calculation amount of this technology is huge, especially the 3-dimensions pre-stack depth migration facing the challenge of the high-dimensional nature of the data. With the development of the parallel computing technology which is based on the cluster-type parallel machine, the solution to the "curse of dimensionality" has been presented to facilitate the new development of the seismic exploration processing technology. In the recent years, the 3-dimensions wave equation pre-stack depth migration has been applied for the volcanic rocks of Daqing Oilfield and ancient buried hill of Shengli Oilfield, which has obtained great progress and provided the technological reserve for the complex structures. However, the speed analysis and amplitude keeping of pre-stack depth migration still need to be solved.

The analysis and processing of the deep data are always the hotspots of the domestic and overseas geophysics research. The progress has been achieved in the complex structure imaging, high solution and weak signal extraction, complex structure modeling, compress multiple-reflection, and AVO, AVA crack show and analysis under complex conditions. The development of the edge-preserving filtering and coherence cube technology has pushed the seismic explanation technology forward and made more accurate geological entity. The trend is the globalization of the collection, processing and explanation. These are the technological support of the continental basin depth and marine pre-cenozoic basin oil and gas resources.

In the development of the oil and gas exploration, the progress of the instruments is very important. The seismic reflection, for instance, has developed from the 1950s tube optical photographic records seismograph (the dynamic scope 25dB), to 1960s analogue magnetic tape recorders seismograph based on the semiconductor devices(the dynamic scope 45dB), then to the 1970s IC digital tape recording seismograph (the dynamic scope 60dB); to 1980s 24-bit analog digital conversion, instantaneous floating-point gain and other systems (the dynamic scope 80dB); and the telemetry system with thousands of channels of application data transmission technology. The progress of the instruments and observation methods has made the seismic reflection as the support of the oil and gas exploration technology. Recently, the high-density single-point digital seismograph development would bring another new reform for the seismic exploration. Therefore, the progress of the instruments and equipments is a necessary part of the technological progress.

The non-shock technology, like the heavy, magnetic and electric, is also the same case. The primary non-shock technology was mainly for the delineation of the basins. With the gravimeter from mGal to μGal, 2-magnitude improvement of the magnetometer accuracy, the non-shock technique started to show the development trend of quantification of qualitative interpretation (like the pattern recognition of the density interface), the elaboration and integration of the quantitative interpretation (like using the seismic data to restrain the Cenozoic impact and using the neural network to inverse the deep interface). Moreover, the real inspection also appeared effect, for example, using the non-shock method to define the igneous rock and fault, separate the base effect and directly define the favorable oil and gas regions. For the electromagnetic aspect, according to the mathematical correspondence between the elastic wave propagation equation and magnetic diffusion equations, the advanced methods and technologies of the seismology would be introduced. Based on the concept of reverse-time migration, the method of reversely extrapolating the seismic wave field from the surface to the underground for the migration imaging was used to explain the electromagnetic field; the study of the reverse-time migration imaging of the earth electromagnetic field was carried out; the offset EMF concept was proposed; and the offset and reversion were combined. In addition, the non-shock exploration technologies like the heavy, magnetic and electric (even covering the remote sensing, nuclear etc.) and processing technology achieved great progress, which were closely connected with the development of the instruments.

Moreover, for meet the higher needs of the current oil and gas exploration, some geological theories and methods of high-accuracy seismic information explanation and analysis (like the basin analysis, sequence stratigraphy, oil and gas system theory, dynamics etc.) were deeply developed, which played important role in new reserve discovery and oilfield production instruction. These new theories and methods can help us dynamically reconstruct the basin sediment-structure evaluation, oil and gas migration, accumulation, deformation process of different-size oil and gas structures and impacts on the oil and gas accumulation and migration etc., reasonably instruct the seismic exploration, reveal the oil and gas accumulation rules and offer service for the oil and gas exploration.

The above mentioned shows that the progress of the oil and gas industry is closely connected with the oil geological theories and technologies and equipments development. As our geology has many different characteristics from the overseas, like the complex structure, frequent structure movements, complex deformation and low permeability of the continental reserves, so how to formulate the new theory, technology and equipments for our own complex geological conditions and guarantee the sustainable development of our oil and gas industry is the vital issue which needs to be solved. The discovery of Puguang Gasfield is a good example.

From the comprehensive study of geology and geophysics, the marine and continental structure frameworks in China are the "three latitudinal strips, two longitudinal strips and two triangles" (Fig. 1-11) [51]. Sichuan basin, same as Ordos basin, is a relatively stable basin which is a superimposed basin with center of the marine sediment. After 70-year exploration, the maximum proved gas reserve was only 40.9 billion m^3 in 2000. The exploration in the deep-water region where was considered to have no growth and no exploration value must break the existing acknowledge of the regional structure-sediment layout. With the instruction of the "region restrains the local and the deep controls the shallow", firstly detect the Permian Changxing Formation standard fossil combination under the defined lower Triassic, re-formulate the layered strata and deny the strata basis of the trough existence. Based on the regional structure background research, deny the trough-growth structure and paleogeographic background, discover many shallow sediment marks in the defined deep-water "trough" area, confirm aurora, the study region, as the Sinian-Triassic Yangtze block in the passive continental margin tectonic evolution stage which should locate in the favorable zone of reserve growth. This all fundamentally resolve the direction issue of the exploration.

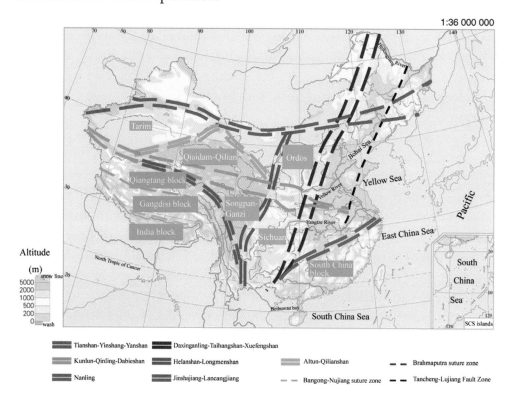

Fig. 1-11   China tectonic frameworks[51]

Three Latitudinal Strips, Two Longitudinal Strips and Two Triangles

The "three latitudinal strips" refer to the straps among the blocks (Huabei, Yangzi, Huanan, and Tarim) in China mainland, i.e. Tian shan-Yinshan-Yanshan, Kunlun-Qinling-Dabie and Nanling.

The "two longitudinal strips" refer to the dense gradient zones in the gravitational fields, which show the drastic changes of China mainland crustal thickness, i.e. Helanshan-Longmenshan, Daxinganling-Taihangshan-Xuefengshan.

The "two triangles" refer to Qaidam-Qilianshan and Songpan-Ganzi areas, which were strongly compressed and reconstructed in the uplifting of Tibetan Plateau.

"Three latitudinal strips, two longitudinal strips and two triangles" is the macro characteristics of China tectonic frameworks, which has the instructive significance for the oil and gas and metal beds exploration.

Meanwhile, the high-accuracy seismic acquisition development has improved the master frequency form 30Hz to 50Hz, and dramatically enhanced the data quality; the study on the mountain high-resolution seismic imaging technology has increased the forecast resolution of the 5,000m and below from 37m to 12m; the establishment of the 5-type seismic recognition models of reef reservoir have achieved the 15m mean absolute error of the 5,000m-deep forecast depth, which could make all the test wells successful. Puguang Gasfield has been rapidly, effectively and safely detected in the exploration area defined by the predecessor as the unfavorable area.

Seismic Image-forming Technology

It is one kind of geophysical technology aiming to perform reversion of subsurface structure, which is covered with abundant rays, velocity distribution and its elastic parameter based upon the record of seismic wave going-through stratum.

From 2004, Puguang Gasfield has expanded proven gas reserve by more than 100 billion m^3 annually for three successive years (in 2004, increasing 114.3 billion m^3; in 2005, increasing 136.7 billion m^3, and in 2006, increasing 105 billion m^3), taking up 20–25% of the yearly total incremental proved gas reserve in national wide.

By the end of 2007, Puguang Gasfield had ascertained gas deposit up to 450 billion, surpassing the reserve volume of Kela 2 gas field—the main gas field of West-East Natural Gas Transmission Project. The area of one favourable facies belt is up to 20,000km^2, equivalent to the area of Shengli Oil Field. It is safe to say that the discovery of Puguang Gasfield is a turning point for oil and gas prospecting theory and pratice, also is a successful example. The formed theory and practice has important guidance function for the oil and gas prospecting of marine facies field of 4.5 million km^2 in our country.

1.4.4 The role and position of oil and gas science and technology development in oil and gas resource field

From the above mentioned development history of world and national oil and gas industry, it can be observed that the progress of oil and gas geology theory and oil and gas science and technology has spurred the development of petroleum industry. It can be seen from the following aspects:

1) The oil and gas science and technology development has deepened people's recognition to world petroleum resource.

With the development of oil and gas geology theory and the advance of exploration and development technology, people's estimation of ultimate oil and gas resource volume has been increasing continuously. For example, the evaluation result of world oil and gas resource of USGS[11] in 2000 shows that global crude oil conventional recoverable resource has reached 411.2 billion t, which has increased by 32% compared to evaluation result in 1994. Take another example of our country: oil and gas resource evaluation in national wide has been organized for three times in history. Due to current cognition degree and influence of oil and gas science and technology, there is a great difference between different evaluation results. The first evaluation result of oil and gas resource was launched in the eighties of twenty century. The result shows that the national petroleum resource volume is 78.7 billion t and the natural gas resource volume is 34 trillion m^3. In the early ninties, the second round of oil and gas resource evaluation result shows that national petroleum resource amount is 94 billion t, and the natural gas resource amount is 38 trillion m^3. The 2003 national oil and gas resource evaluation shows that our national prospective petroleum resource amount is 108.6 billion t, geological resource amount is 76.5 billion t, recoverable resource amount is 21.2 t, the prospection is at medium stage; the prospective natural gas amount is 56 trillion m^3, the geological resource volume is 35 trillion m^3, the recoverable resource amount is 22 trillion m^3, the prospection is at early stage; coal-bed gas geological resource volume is 37 trillion m^3, the recoverable resource amount is 11 trillion m^3; based upon the shale oil converted from oil-shale, the geological resource volume is 47.6 billion t, recoverable shale oil is 12 billion t; bitumen geological resource volume is 6 billion t, recoverable resource amount is 2.3 billion t. Apart from the conventional oil and gas resource evaluation, the new round of oil and gas also launched evaluation for unconventional oil and gas resource, such as coal-bed gas, shale oil, and bitumen. The process of three resource evaluations demonstrates that people's cognition to oil and gas resource changes with the development of oil and gas science and technology.

2) The development of oil and gas science and technology has flourished the exploration and development of oil and gas resource.

The development of petroleum geology theory and the advancement of science and technology have improved human's ability to explore the oil and gas resource substantially. Although during the twenty years from 1985 to 2005, the global accumulated recovered petroleum has surpassed 70 billion t, owning to the progress of science and technology, more resource has been explored continuously and the global petroleum remaining proven reserve has been increasing everlastingly, in 2005, the remaining proven reserve reaching 164.5 billion t, which has increased by 35.7% [52] compared to 105.5 billion t in 1985. At the same time, the progress of science and technology has stimulated the development of oil and gas exploartion. The emergence of reflection seismics, rotary drilling machine, plate tectonic theory, kerogen hydrocarbon generation theory, waterflood development, tertiary oil recovery and 3D seismics has helped to promote the global petroleum production continuously (see Fig. 1-12). Due to the development of science and technology flourishing the oil and gas exploration and development, which results in the increases of global petroleum remaining recoverable reserve and petroleum output, the predicted global petroleum peak by some organizations and experts has been put off time and time again.

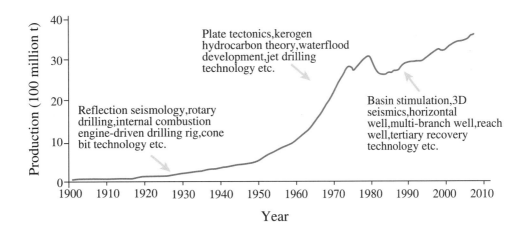

Fig. 1-12   The science and technology development bring forth the leapfrog development in global petroleum production (according to Yanqing He[52], 2006, revision)

3) The development of oil and gas science and technology improving the cost benefit of exploration and development.

It is the goal of all the oil and gas companies to obtain more oil and gas reserve and output with lower input, also it is the basic driving force of global oil and gas science and technology development. In the nineties of twenty century, when the oil price was low, relying upon technology advance, investment and core business management, the international major oil companies managed to bring the cost of exploration and development down by 33%. In recent years, the emergency and appliance of 3D seismics technology, DLWD technology, geo-steering drilling technology, and tertiary oil recovery brought up the drilling success ration substantially, and accelerated the drilling speed. Per well cumulative production also increased greatly. Although from the perspective of exploration cost, it is on the ascendant trend. This is caused by many aspects. Due to the rise of international oil price, every oil and gas company made adjudgement in the cost strategy in order to gain maximum benefits. This is a very important cause. But from the perspective of oil and gas products cost components, the ration of direct investment used for exploration in the total investment in entire oil and gas production is shrinking, and the cost benefit is increasing on the whole.

4) The development of oil and gas science and technology expanding the oil and gas resource exploration field .

With the development of oil and gas science and technology, and the progress of instrument and equipment technology, oil and gas exploration field has been expanded continuously. With the advance of offshore seismic, drilling, oil recovery, oceanics industries equipment and technology, the oil and gas exploration expanded from land to ocean. At present, offshore oil and gas output has taken up about 8% of the global total output, and it is on the ascendant trend. Besides, the development of oil and gas science and technology and offshore industries equipment technology enabled the offshore oil and gas exploration to move from shallow water to deep water. For example, Petrobras has discovered Tupi oil field at the Santos Basin. The oil field is expected to have 0.68–1.1 billion t reserve volume[53]. Its main reservoir—saline facies carbonate rock reservoir bed is nearly 7,000m [54] from the ocean surface. At present, operating water depth of offshore oil and gas exploration has exceeded 2,000m, and the operating area has expanded from the North Sea and Gulf of Mexico to West Africa, South America and Australia continental shelf. Besides, the expansion of oil and gas exploration field can be seen from the fact that the exploration target has moved from shallow position to deep position. At present, the drilling depth of land oil and gas exploration has exceeded 8,000m. What's more, it is included in many countries and companies' plan to explore oil and gas resource in polar area and region with complicated surface environment.

5) The development of oil and gas science and technology promoting the clean development of petroleum industry.

The exploration and utilization of oil and gas resource brings convenience to human's life and has made undeniable contribution to the global economic and social development, but at the same tine, it causes many environmental problems. The development of oil and gas science and technology has brought forth tremendous environmental benefits, imposing less influence to the environment during oil and gas exploration, and meeting human's stricter environmental requirements. For example, the drilling technology development of direction well, horizontal well, post hole, extended reach well, branch well, cluster well makes the area occupied for well site become smaller and smaller and reduces the environmental pollution greatly. Besides, the application of advanced technology of oil field produced water treatment reduces the discharge of pollutants during the process of oil field exploration. At present, combination of advanced technology with HES management system (Health Safety and Enviroment Management System), can reduce the discharge of pollutions during the process of oil field exploration, bring down the environmental pollution degree gradually and increase the safety degree of operators.

6) The development of oil and gas science and technology strengthening the sustainable development capability of petroleum companies.

The development of oil and gas science and technology strengthens the sustainable development capability of petroleum companies and the capability to cope with the market competition. The progress of oil and gas science and technology not only increases the potential recoverable oil and gas resources volume, expands the exploration field, but also promotes the cost benefits of petroleum companies, making the company take the advantaged position in the competition with other non-petroleum energy companies. Some key technologies, particularly those imposing a significant influence to the present and future oil and gas exploration, such as four-dimensional seismic, oil reservoir integrated description, intelligent well, numeric oil field and deep water exploration, have become the main driving force to propel the increase of present and future oil and gas reserve volume and output and guarantee the sustainable development of company.

1.5 The trend of science and technology development in oil and gas resource field

Facing with the increasingly fierce oil and gas supply-demand situation, every country and every petroleum company in the world is acting positively to propel the advancement of science and technology in oil and gas field and striving to take an advantaged position in the future competition. The main development trends are as following:

1) Individual technology shifting to comprehensive technology, the multidisciplinary combination and technology integration emerging gradually.

With the increase of demand of exploration and production for comprehensive service, the development of many engineering technology is finished by the cooperation of different specialist personnel. The projects and products developed by the cooperation of different specialist technology center are increasing. For example, geo-steering drilling requires the collaboration of personnel of drilling, logging, geology and information technology majors; the hydrocarbon source comprehensive forecasting techniques require the participation of personnel of geochemistry, geology and deposit majors; what's more, reservoir real time management technology, numerical oil field technology, intelligent oil field technology (smart field) [56] require the cooperation of personnel of different majors. As the future resource and exploration target evaluation integrative technology, it is scientific and multiaspect (resource, geology, engineering, economy and risk) evaluation of drilling target, and provides support for the investment decision-making of oil and gas companies. This highlights the combination of multidisciplinary and intergration of technology.

2) The combination of innovative and high technology with traditional oil and gas exploration technology, reformation and industry structure optimization.

In twenty-first century, science and technology revolution centers on biological science, information science, material science and cognitive science has imposed significant influence to the oil and gas industry. The development and wide application of information technology has brought great impact to many traditional industries. At present, the information technology is changing the traditional oil and gas exploration technology fundamentally. Down hole information (all the data concerning oil and gas reservoir and collected from wellbore) is collected and transmitted onto the ground by various sensors and instruments. The transmitted data can help the related experts make decision and help customers shorten the decision-making period, optimizing the oil and gas reservoir exploration and obtaining more economic benefits. Therefore, some famous foreign petroleum companies put an emphasis on the enterprise information construction, not only establishing network system in world wide, but also developing abundant application system supporting information share and real-time decision-making to guarantee the cooperation of technology development and knowledge share and quick response to the market demand. For example, Schlumberger, the largest oil field technology service company in the world, has set up branch companies and research department in many oil and gas exploration countries. Information share and organization of experts in different places to treat and explain [57] information collaboratively can be implemented by international network. Besides, with the application of new findings of biological science, material science and cognitive science in oil and gas industry, the following technologies can be realized in the future, such as microbial oil production strain DNA reformation, oil displacement polymer molecule design, nanometer oil production equipment and intelligent drilling.

3) Emphasizing the economic and environmental benefits of technology and taking the path of technology virtuous circle and sustainable development.

Foreign government and petroleum companies put a great emphasis on the commercial use of technical results. The ultimate purpose is to apply the technology effectively and bring forth economic benefits. At the same time, with more attention on the environmental problems, the environmental problems in the process of oil and gas exploration has attracted more and more attention from the governments and petroleum companies, and the environmental problem has become one of the indicates of technology progress. For example, the post hole drilling technology, laser drilling technology, CO_2 flooding oil production technology exhibit this trend. What's more, the technology innovation process is greatly accelerated through close combination of developed project with production, customers' demand-orientation and close cooperation with customers.

4) Scrambling for oil and gas resource in public area has become an important feature in future oil and gas exploration in the world.

Since abundant oil and gas resource has been found in public area, such as North Pole, Caspian Sea and Western Pacific marginal sea or area whose escheatage is in dispute, the sovereignty dispute has never stopped. Since Russia planted their national flag under the Arctic Ocean, this competition becomes more and more fierce. Russia, Canada, Norway, Denmark, America are acting positively to explore the North Pole resource, and they even acclaimed their sovereignty to North Pole area from United States. In order to take a lead in exploring the oil and gas resource in Arctic Ocean, Russia has resumed strategic cruise above the Arctic Ocean. Canada has organized joint maneuver inside the artctic circle. United States Geological Survey has organized evaluation[58] of oil and gas resource in North Pole area.

5) The strengthening of unconventional oil and gas resource exploration has become one of important means to guarantee future oil and gas supply for countries and oil and gas companies in the world.

Oil and gas resource is one kind of non-renewable resources. With the continuous exploration of oil and gas resource, the conventional oil and gas definitely will be exhausted some day. Therefore, in order to meet human's demand for oil and gas resource, many countries and petroleum companies has considered exploration of unconventional oil and gas resource, such as oil sand, heavy oil, coal-bed gas, gas hydrate, oil shale and water-soluble gas, as an important means to guarantee future oil and gas supply. For example, America, Japan and India once stipulated national gas hydrate R&D plan, and settled schedule for commercial exploration of hydrate. Besides, America organized resource evaluation for gas hydrate in the Alaska North Slope in 2008[62]. Take another example, some world famous petroleum companies, such as ExxonMobil, Chevron and Shell, have set up Unconventional Oil & Gas Department, strengthening the exploration of unconventional oil & gas[55, 63].

Gas Hydrate

Gas hydrate is a kind of non-stoicheometric and ice-like crystal clathrate formed by water and natural gas under certain conditions (suitable temperature, pressure, gas saturation, water salinity and pH value). Since it looks like ice and will burn when catching fire, it is also called "flammable ice" or "solid gas" and "gas ice". The compositions of natural gas, such as CH_4, C_2H_6, C_3H_8, C_4H_{10} and CO_2, N_2, H_2S can form single or multi-gas hydrate. The main composition of gas hydrate is methane. For the gas hydrate with more than 99% of methane, it is usually called Methane Hydrate.

6) Government, petroleum company and technology service company working together to promote the progress of oil and gas science and technology.

In the development history of oil and gas science and technology, governments, petroleum companies and technology service companies have played an important role and made great contribution to the R&D and application of oil and gas science and technology. At present, there is a trend in many countries that governments, petroleum companies and technology service companies working together in R&D of oil and gas science and technology. Governments are the leader, organizer and investor in oil and gas basis theory study; petroleum companies are investor, developer and user of oil and gas science and technology; technology service companies are developer, spreader and supplier of oil and gas science and technology.

2 Main Fields of Oil and Gas Exploration and Problems

2.1 Continental basin

With the strengthening of oil and gas exploration, the more and more complicated prospecting targets and the more deeper the more complicated of structure's controlling effects on oil and gas basin formation, it can be shown from lithostratigraphic trap increasing, the ratio of unconventional reservoir increasing, prospecting target becoming smaller, and the ratio of deep targets increasing. It is urgent to solve these problems with new knowledge and technology; at the same time, technology reserve and basic material for developed basins can hardly satisfy with the present demand. Seismic acquisition configuration and wellbore test is relatively out of date, and it fails to make a systematically comprehensive cognition and exploration of complicated block fault and lithologic hydrocarbon reservoir, and limits the deepening of maturing oil field exploration. Therefore, it is urgent to deepen the study of continental basin remaining resource potentiality and petroleum geology rules. The main prospecting target for continental basin is concentrated on the basin litho-stratigraphical hydrocarbon reservoirs, low porosity-low permeability reservoirs and deep reservoir.

2.1.1 Deep oil and gas exploration

After more than fifty years' exploration, the course of seeking shallow, high abundance and large-scale new reservoirs in continental basins in China has came to an end. The main oil fields in the eastern China have entered into "high extraction, high water cut" phase. In 2005, the oil field composite water cut exceeded 80%. The distribution of land remaining oil and gas resource in our country is uneven. Generally, 39% of remaining petroleum resource and 57% of remaining natural gas resource is located at the deep zone. In this book, "deep zone" prospecting target roughly refers to layer system under the present basin main prospecting target interval. For eastern land basin, it generally refers to depth scope of within 4,000m; for western land basin, it refers to depth scope of within 5,500m.

The litho-stratigraphical hydrocarbon reservoir has taken up more than 60% of our new increased reserve, and most of them are low permeability, even ultra-low permeability reservoir. In the litho-stratigraphical reservoir exploration, we are now confronted with new scientific problems and technical bottleneck, such as heavy sandstone trap validity assessment, large area low-permeability lithologic hydrocarbon enrichment region prediction, large-area marine facies sandstone formation oil and gas distribution, and reservoir prediction.

In recent years, the important breakthrough in deep oil and gas reservoirs exploration has been made successively in Bohai Sea Gulf, Sichuan and Tarim Basin. A good exploration prospect within 7,000m is shown in the western basin. The prospecting targets are on a trend of diversity and complication. Volcanic reservoir rocks will become main prospecting target and exploration field in future.

The trends mentioned above are turning the prospecting targets to the deep reservoir. Therefore, a series of theory and technical problems, which are limiting the deep reservoir exploration, are caused. Oil and gas bearing basins in our country have experienced multi-phase Tectonic activities. The original deep reservoir has been changed greatly, and we are confronted with unprecedented difficulties in basin deep reservoir exploration theory and technology. The development of oil and gas exploration is directly confined. Former experience and domestic and foreign current theories can hardly be directly used in deep reservoir exploration. Therefore, it is very important to recognize the conditions of deep oil and gas formation and reservoir rule. The key technology in deep reservoir exploration will become a key issue.

At present, the following aspects in deep reservoir exploration require urgently to be solved:

1) Deep basin formation and reconstruction.

Archetypal basin restoration has significant meaning for recognition of the spatial layout of favorable depositional facies in the process of basin formation , the position and scale of hydrocarbon source and reservoir rocks, after multiple basin superimpositions, structural activity and complicated reformation. But the complication of the deep structure of basins in China and intensity of later strutural activity caused a series of difficulties in archetypal basin restoration and the determination of key reformation phase.

The development of high quality of source and reservoir rocks and their superimposed relationship are mainly constrained by structure-palaeogeography. The filled tectonic framework for archetypal basin restoration and its evolution are the main aspect and basis of this study. For the basin deep structure, the following problems urgently require a deepened study in the development environments of deep hydrocarbon source rock, reservoir and cap rock: how to set up isochronous stratigraphic frameworks, restoration of primary basin-mountain relationship and basin boundary from the perspective of activity, recognition of the rise and fall of sea surface's determination effect on favorable facies belt, reestablishment of paleoclimate changing history.

The prospecting layers in the deep basins have usually experienced multi tectonic reworking, featuring structural superposition reformation. The following problems have become key and difficult issues in archetypal basin restoration and determination of main reservoir-forming stage: how to recognize main deformation of deep basin rock formation through deformation analysis, to strengthen the study of basin-mountain structure evolution, and dynamically reflect multiple evolution and post superposition deformation history of deep archetypal basin.

2) Evaluation of high mature hydrocarbon source in deep basin.

Evaluation of hydrocarbon generation from high mature source rock is one of important unsettled problems in deep oil and gas exploration. The higher maturity, the more unanimous the organic matter nature is. It is very important to recognize the difference between high mature organic matter and its nature for the recognition of main hydrocarbon source rock, understanding the process of hydrocarbon generation from source rocks in the geological history. It is failed for the methods of organic carbon abundance and quality of source rocks, and current static maturity to evaluate the hydrocarbon generation of high mature source rock. It is necessary to develop dynamic evaluation method. That is to say the following problems require to be settled down urgently for the resource evaluation in the process of deep oil and gas exploration and favourable prospecting target preference: describe quantitatively the process of hydrocarbon generation from source rocks during geological history or higher thermal maturity, and concrete determination of hydrocarbon generation period. For recent ten years, the establishment of new hydrocarbon generation kinetics and thermal history has laid a foundation for dynamic evaluation of hydrocarbon source. Another important problem confronting the hydrocarbon generation evaluation in deep basin is the non-pyrolytic reaction, which is oxidation reaction of organic matter, could influence the capability of hydrocarbon generation from high mature source rock greatly. But not enough attention was paid to such reaction until present. It is important to understand the terminal of hydrocarbon generation from source rock, and CO_2 enriched reservoir.

Evaluation of hydrocarbon loading capability and hydrocarbon release capability for high mature source rock is a primary problem requiring to be settled in deep shale gas exploration. There is great difference in physical nature between organic matters of different types and different evolution stages, which result in the difference of absorbing and desorption capability. It is a critical issue in high mature shale gas about how much natural gas can be absorbed in the certain volume of hydrocarbon source rock, and how much natural gas can be released under different geological conditions. At present, such study has not been done much in the international field. But with the improvement of shale gas exploration, it will definitely become a hot issue.

The determination of deep hydrocarbon source kitchen (hydrocarbon and non hydrocarbon source rocks) is a difficult problem in source rock evaluation. It is very important for deep basin exploration to recognize hydrocarbon source kitchen's contribution to reservoir, indicate the hydrocarbon compound features at different stages with various kinds of means, and identify and determine major hydrocarbon source rock under the condition of coexistence of different kinds of hydrocarbon source rocks. Particularly at the deep part with little drilling and geophysical material, a new technical path of hydrocarbon source kitchen must be discovered. Inversion should be adopted as major determination technology.

3) Formation and forecasting of effective reservoir in deep basin.

From the recent exploration, it can be seen that the high quality of sandstone reservoir within 4,500m depth and carbonate reservoir with high porosity and permeability in 7,000–8,000m can still be discovered in the deep basins, suggesting that the quality reservoir of clastic and carbonate rocks in deep basin exists and the prospect for oil and gas exploration in deep basin is promising. But since the dominative factor and distribution of quality reservoir has not been figured out, the oil and gas exploration in deep basin cannot take a further step. Besides, the neopaleozoic volcanic reservoir in middle and west China also has a good prospect for exploration.

At present, formation and evolution of volcanic rock (such kind of new reservoir) and development of porosity and permeability in deep basin is still at the early stage. The following problems have become the exploration difficulties: combining reservoir of volcanic rock reservoir-embedded diagenetic facies, process of later structure-fluid medium superposition reformation and its influence to fracture-cave development distribution.

4) Formation of oil and gas pool and favourable exploration area in deep basin.

Comparing with middle and shallow oil and gas pools, the geological environment, temperature and pressure conditions, physiochemical nature of matter in deep basin has undergone great change, causing the low permeability or/and ultra-low permeability for clastic reservoir. Based upon current knowledge of reservoir formation, the reservoir with the condition of low permeability and ultra-low permeability can hardly be formed, and the favourable conditions can hardly be satisfied in the natural basin environment. In recent years, some low permeability and ultra-low permeability reservoirs have been discovered in deep basin. But the progress of exploration and development is blocked, since the mechanism of reservoir formation and distribution has not been clearly figured out. The carbonate rock in deep basin has been undergone multiphase structure superposition, and reservoir formation period featured multiphase adjustment reformation. The critical part is complex multivariant transforming system dominated by multiphase fractures. However, since the lack of research on formation process of carbonate rock transmitting system in deep basin, distribution feature and hydrocarbon migration-accumulation efficiency, the hydrocarbon distribution in carbonate rock reservoir of deep basin has not been figured out.

It is very important for the research of complex transmitting system in major structure evolution stages and major reservoir formation period, fluid medium connectivity feature and aeolotropism of passage system, structure geological effects and factors in deep basin to understand oil and gas reservoir formation and distribution, based upon fluid medium feature under high temperature and high pressure and the relationship between fluid medium and rock.

At present, there are many evaluation ways for favourable exploration area. Statistical method is the basis. These methods for the middle and shallow oil and gas exploration area with more information is adequate. But for those deep basins with low extent of exploration, and poor information, those methods can hardly be used, which constrains the selection of exploration direction and evaluation of favourable target in deep basin. It is urgently required to develop a new applicable evaluation technology and method for oil and gas reservoir formation and distribution in deep basin.

Application study of geological modeling, non-seismic and seismic detection key technology and reservoir forecasting in deep basin above mentioned should be launched based upon reservoir development and distribution pattern to improve deep detection resolving ability and accuracy. This can provide important support for understanding of reservoir formation in deep basin, developing key technology and forecasting favourable target.

2.1.2 Litho-stratigraphical oil and gas reservoir

For recent ten years, the number of the litho-stratigraphical hydrocarbon reservoirs take up an increasing. For the North America region, where the degree of exploration is high, litho-stratigraphical reservoir takes up 40% of total discovered number. The statistics of global hydrocarbon resource show that amount of hydrocarbon resource in litho-stratigraphical trap is roughly equal to structural trap [64]. With the progress of exploration technology, the importance of litho-stratigraphical reservoir will be manifest.

litho-stratigraphical reservoir

The closed hydrocarbon low energy position with lithologic changes around reservoir body or in updip direction, or with impermeable lithologic stratigraphy with absence, denudation or overlap of stratigraphic sequence is called stratigraphical trap. As the accumulation of industrial scale hydrocarbons, it is called sratigraphical hydrocarbon reservoir. Trap formed due to lithologic change of reservoirs is called lithologic trap. After the accumulation of industrial scale hydrocarbons, it is called litho-stratigraphical oil and gas reservoir.

In our country, since all main hydrocarbon bearing basins are continental facies depositional basin, it is very favorable for the formation of litho-stratigraphical reservoir and its preservation under conditions of complex and variant lithostratigraphy and lithologic facies. Since nineties of twenty century, with increase of litho-stratigraphical reservoir exploration, application of sequence stratigraphy method and enlargement of high resolution 3D seismic acquisition scope, the achievement of litho-stratigraphical reservoir exploration has been significantly gotten. Middle and large size litho-stratigraphical has been discovered in basins. Thus, the proven oil and gas reserve has been increasing continuously and the litho-stratigraphical reservoir reserve has taken up about 50% of total discovered amount. This has significant meaning for the security of stable hydrocarbon production growth. In a long term, such reservoir will guarantee the growth of oil and gas reserve from exploration and production.

But we should know that great difficulties still remain in litho-stratigraphical oil and gas reservoir exploration. The most serious problem is the short of geological, geophysical comprehensive forecasting method and technology to identify and prescribe litho-stratigraphical reservoir. The main problem is high accuracy forecasting of favorable reservoir under deep complex structure and stratigraphical background. Especially little was known about central and western litho-stratigraphical reservoir distribution. This can be seen that sedimentary facies mode requires improvement, relationship between deep fracture and litho-stratigraphical reservoir is unclear, favorable reservoir and distribution has not be mastered, mechanism of litho-stratigraphical trap formation is unclear, reservoir formation of hydrocarbons from source outside and mechanism of hydrocarbon generation kinetics is poor; For the formation and whole distribution pattern of marine sandstone and litho-stratigraphical reservoir, little is known.

2.1.3 Low porosity, low permeability oil and gas reservoir

With the improvement of exploration and development of technology, more attention is paid to low permeability and/or ultra-low permeability hydrocarbon resource exploration and development. This phenomenon is very prominent in China. Low permeability reservoir has the following features: narrow sediment, low maturity of sediment structure and mineral, poor physical property of reservoir, small pore throat radius, low matrix permeability, big diagenetic difference, high stress sensitivity, split developed and strong non homogeneous in macro and micro aspect [65].

According to Chinese clastic rock reservoir physical property evaluation standard [66, 67], if reservoir bed permeability is lower than $50 \times 10^{-3} \mu m^2$, it can be called as low permeability reservoir. In recent years, 70% increment of hydrocarbon reservoir in the continent hydrocarbon bearing basins, China is low permeability reservoir, among which specially low permeability reservoir with permeability lower than $10 \times 10^{-3} \mu m^2$ takes up 50%.

Since formation mechanism of low-porosity and low-permeability reservoir and complex distribution pattern, at present, little is known about reservoir formation features and its relationship with micro faulted structure and stratigraphical distribution. There is great difficult in exploration. Progress of exploration and development of low-permeability reservoir is influenced. In Northeast basin, Ordos basin, Zhunger basin and Qaidam basin, the low porosity, low permeability hydrocarbon resource takes up main residual proven reserve. Reservoir formation mechanism, distribution pattern of such reservoir and exploration technology are different from those of general litho-stratigraphical reservoir.Attention should be paid to its reservoir formation process and identification technology, including the following aspects:

1) Geological structure and microcrack formation mechanism and its controlling effects on porosity and permeability development of oil and gas reservoir.

2) Relationship between pressure and reservoir space.

3) Basin-Mountain coupling process, fluid medium, relationship between magmatic rock activity and reservoir.

4) Low porosity, low permeability reservoir numerical simulation (comprehensive simulation of gravity, fluid pressure and stress).

2.1.4 Superposition basin exploration

According to estimation, about two thirds of hydrocarbon residual resource in China is distributed on the land. 80% of the resource is distributed in superposition basin. Multiphase evolution in superposition basin caused superposition development of lower Paleozoic carbonate rock, upper Paleozoic/Mesozoic coal measures, Meso-cenozoic lake hydrocarbon source rocks. During the process of shallow continental oil and gas exploration in the superposition basin, the achieved continental oil generation theory and a series of effective exploration technologies is established. In contrast, exploration theory and technology for deep marine hydrocarbon in middle and low parts of superposition basin is still being explored.

The current exploration shows that in eastern maturing field, ascertain degree of main bed of interest is high. The deep hydrocarbons will be the main field for future exploration, especially Precenozoic layers; at western area with low exploration extent, the potential of hydrocarbon resource is great, hydrocarbon resource to be explored is also mainly contained in Precenozoic layers. The cost of exploration is high, the success ratio of drilling is low, and the increment of hydrocarbon reserve is slow. Therefore, the deep oil and gas exploration in superposition basins encounters unprecedented difficulties in oil geological theory and hydrocarbon exploration technology. The main problem lies in that little is known about hydrocarbon generation-release mechanism, hydrocarbon generation peak of source rock and reservoir formation period; the knowledge on the II-zone hydrocarbon potential derived from multi-stage tectonic superposition or new tectonic effect, and distribution of superimposed oil and gas reservoir through multi-stage tectonic changes and adjustments of different strata of oil and gas and other issues is limited;little is known about foreland basin nature, structure, structure evolution, depositional environment, and sedimentary sequence, and the hydrocarbon generation kinetics system, reservoir cap assembly, oil and gas accumulation conditions under above background. Studies should be launched about structural evolution sequence, superposition method and its influence to reservoir formation, adjustment and reservoir history.

Key scientific and technical problems for future continental oil and gas exploration including:

1) Global structural activity to constrain the type of classical superposition basin and its genesis in China.

2)Analysis of deep complex structure features, deep high quality reservoir development and hydrocarbon generation kinetic mechanism, reservoir formation history and forecasting technology.

3) Reservoir formation process and distribution in complex structural belt, influence of deep complex fracture to oil and gas reservoir under extreme temperature and pressure, and research on formation mechanism of lithostratigraphy oil and gas reservoir and technology of precise detection.

4) Geophysical imaging technology under complex earth surface and complex structure background.

2.2 Precenozoic marine carbonate rock layer system

Compared with foreign marine layer system, the Precenozoic marine strata in China have undergone reformation of complex structure for many times due to long history, and mostly, taken on in the form of marine residual basin. The universal features of Precenozoic marine reservoir are: deep buried, fracture system complexity, type multiform, various oil sources, and scattered distribution and highly elusive. This becomes tremendous difficulty for exploration. From the previous research achievement and exploration results, it can be observed that precenozoic marine residual basin reservoir can take on in the form of paleo burial hill, bioherm, beachrock, and dolomite, weathered crust. This forms are various and complex. At present, this research on its formation mechanism, relationship and spatial layout is not enough. What's more, complex surface, complex medium, complex structure, complex reservoir has brought forth a series of technical problems, which directly constrains the precenozoic marine exploration. Puguang marine reservoir and others have proved that main reservoir is located at offshore high-energy facies of continental margin. Therefore, the research of paleoclimate, paleoenvironment with the target of seeking old coast belt and high-energy facies belt from the perspective of global paleostructure and paleoenvironment has been increasingly paid a attention. What's most important, while great breakthrough was made in marine oil and gas exploration, it is also paid more attention to the dolomite rock to precenozoic marine oil and gas contribution in. From the present research achievement, we can see what carbonate rock, neither hydrocarbon source rock nor quality reservoir, has made important contribution to marine reservoir formation, is close related with dolomitic reservoir. The liquid hydrocarbons discovered at depth of 8,000m in Tarim Basin also comes from dolomitic reservoir. Although dolomitic mechanism is still in argument, it can be considered that deeply laying hydrothermal dolomitic activity has close tie with later hydrothermal activity (structure reformation). Whether global structure reformation has provided thermal reformation condition for ancient hydrocarbon source rock and influence of structure reformation and magmatic activity to hydrocarbon formation may become a hot frontier research issue.

Marine Residual Basin of Pre-Cenozoic Era

Apart from inland clastic rock sedimentation basin in Pre-Cenozoic Era in China's marine and inland area, the Pre-Cenozoic Era has also enjoyed wide distribution of marine carbonatite. They have the good oil generation conditions like as continental sources; however, large-scale mountain making movement in the Mesozoic Era has led to violent extrusion and modification. Although it has lost the primary basin form under the extrusion of basement separation and overthrust nappe, it has still reserved huge amount of oil and gas resources.

In oil and gas exploration of marine carbonate, various large oil and gas fields in recent years has been discovered in China, such as Lunnan and Tazhong in Tarim Basin, Puguang, Yuanba and Longguang in Sichuan Basin. The newly increased oil and gas reserve in carbonatite layers has constantly increased and it has taken an even important position. Compared with foreign countries, the oil and gas exploration for carbonate layers is confronted with various difficult problems such as big reservoir heterogeneity, great prediction difficulty of storage layer, high maturity of organic matters, multiple reservoir formation periods, deep buried target and complex pressure system. Therefore, we can not have a clear knowledge of the oil and gas reservoir formation and regional rules. Relevant exploration theories and technologies are still in the threshold and development stage.

Since the sedimentary basins have undergone numerous periods of basin formation and modification, complex reservation history and numerous periods of fluid activities, the deeply buried karstification and dolomitisation are very active. The Neopaleozoic violent structure-magmatic movement and relevant heat fluid activity have made the extreme complexity of the deep carbonate diagenesis. It shall launch out the research of multiple structure-fluid activity into the deep carbonate diagenesis and modification and make clear the main control elements of high quality reservoir. It is known as one of the key elements to unveil the distribution of deep oil and gas reservoir.

In order to promote the exploration and development of marine oil and gas resources in Pre-Cenozoic, it is urgent to launch out the basic research including geological conditions of oil and gas reservoir and geophysical probing technologies. It shall include the following contents:

1) Relationship between global scale structure and China's oil and gas.

2) Control of climate evolution to hydrocarbon source rock.

3) Basic knowledge of formation and evolution of Tethyan ocean as well as relationship with oil and gas bearing basin; make clear the dynamic distribution of lithofacies paleography and structural framework controlling basin upheaval and indentation under the guidance of mobilism viewpoint.

4) Accurately resume the structural stress field of marine stratum and its changes, conduct quantitative research into the structural deformation and its influence on the oil and gas transportation, displacement, gathering, storage or damage through establishment and solving of basin geodynamics model.

5) Restriction of regional structure on the basin structure and stratigraphic development features; explore into the formation and evolution process according to relationship between basin and surrounding blocks as well as deep kinetics mechanism.

6) Launch out research into marine oil & gas hydrocarbon generation kinetics and reservoir formation theory, especially dolomitisation mechanism; know the complex process and horizontal change of dolomitisation; establish the reservoir distribution and prediction model, and instruct exploration and deployment of oil and gas reservoir.

7) Develop the comprehensive geophysical technologies applicable to the Pre-Cenozoic marine oil and gas exploration on the basis of in-depth knowledge of structure time & space evolution and reservoir distribution model, unveil deep structure features, predict reservoir distribution and circle favorable areas and belts.

2.3 Deep basin in sea area

According to the incomplete statistics, 1,200–1,300 oil fields in the world have been discovered in the deep water turbidite systems and relevant sedimentary systems or have started formal production. Most of them have enjoyed high production which has taken up 60% of the global increase proved reserve in recent years.

China's South Sea is rich in broad deep water areas with a land area of about 1.20 million km^2 in the land slope deep water area; about tens Cenozoic sedimentation basins are partially or completely located in the deep water area of Northern land slope area of South Sea and Nansha water area. In June 2006, LW3-1-1 deep water exploration well (1,480m) first drilled in Baiyun sag of South Sea Pearl River basin has encountered 73.2m gas layer and made serious breakthrough in natural gas exploration of Northern deep water area of South Sea; in the Southern area of South Sea, Philippines and Malaysia have launched out the development of deep water oil and gas fields. Prospect of deep water oil and gas resources is broad in the South Sea.

However, exploration and research of Mesozoic oil & gas and deep water oil & gas at the edge of South Sea is at a very low level; it is urgent to conduct basic research into the stratigraphic distribution of deep water basin, basin structure features and land edge structural evolution history and also research into the special oil and gas geological features of deep water areas (such as hydrocarbon generation process under abnormal temperature and pressure). Therefore, it can answer a series of problems such as "What is potential and exploration prospect of deep water oil and gas resources? What are favorable oil and gas areas and belts as well as new exploration fields and directions? What is optimal strategic selection area and field of oil and gas resources?" The scientific and technological problems can be summarized as follows:

1) Basic geological features such as structure features and geodynamics background formed by China's deep water basin.

2) Hydrocarbon generation dynamics mechanism of oil and gas under the abnormal temperature and pressure in deep water basin; basin structural evolution; control action of structural deformation in different periods to migration, accumulation, preservation and distribution of oil and gas.

3) Prediction of potential and exploration prospect of oil and gas resources in China's deep water basin.

4) Geophysical exploration, drilling and prospecting equipment and technology series under deep water-ultra deep water conditions.

2.4 Unconventional oil and gas reservoir

With the constant consumption or price increase of traditional fossil fuel resources, the unconventional oil and gas resources have played an important role in the global energy structure and aroused the general attachment of various countries. In 2005, Joint Research Center of European Union released the research report entitled "Prospective Analysis of the Potential Non-conventional World Oil Supply: Tar Sands, Oil Shales and Non-conventional Liquid Fuels from Coal and Gas"[68] to conduct a rather systematic analysis to the reserve development technology and supply prospect of unconventional petroleum resources such as oil sand, oil shale, coal-to-liquids and gas-to-liquids; it has provided the reference for the development of global unconventional petroleum resources. In 2006, United States released the special planning report of unconventional fuels entitled "Development of America's Strategic Unconventional Resources"[69] to push forward the research and exploration of unconventional oil and gas resources in United States.

The unconventional oil and gas resources shall mainly include the coalbed gas, oil sand, oil shale, natural gas hydrate, and shale gas, dissolved gas in water and oil & gas reserved in volcanic reservoir. Globally, the unconventional oil and gas resource is rich at present, the production of global unconventional oil has exceeded $7,500\times10^4$ t/a and production of unconventional natural gas has exceeded $1,800\times10^8$ m^3/a[70]. The unconventional oil and gas resources and unconventional energies in China is also rich, such as oil shale, oil sand, coalbed gas, natural gas hydrate and dissolved gas in water, enjoying enormous development and utilization potentials. According to the new evaluation result of oil and gas resources, coalbed gas has realized a geological resources of 37×10^{12} m^3 volume, shale oil has a geological resources volume of 476×10^8 t, and tar sand oil has a geological resources volume of 60×10^8 t[15].

Since 1990s, China has done huge amount of work in the coalbed gas, oil sand mineral and oil shale, attained important achievements and laid a solid foundation for the large-scale unconventional business. Compared with the research and utilization of world unconventional oil and gas resources, China is rather backward in the research and development of unconventional oil resources and has fewer endeavors in the research into potential research, evaluation technology, exploitation technology and comprehensive utilization technology of unconventional oil and gas resources.

China has launched out the research and exploration of unconventional oil and gas resources which mainly include the coalbed gas, deep basin gas and shale oil. However, the general research degree is rather low. With the price increase of world crude oil, it has started the exploration and development of heavy oil/asphalt and tar sand oil; the basin deep layer gas and shale gas widely explored in United States and Canada have not proceeded substantive work in China yet. Given the limited knowledge and exploration degree of such oil and gas reservoir, we shall first settle the following scientific and technological problems:

1) Resources potential and distribution rule of unconventional oil and gas reservoir.

2) Formation theory, structural deformation of unconventional oil and gas reservoir and relationship between layer and stratum.

3) Prediction and evaluation technology of unconventional oil and gas reservoir.

4) Key technologies to upgrade the yield rate of unconventional oil and gas reservoir.

2.4.1 Coalbed gas

The coalbed gas is one of the hot spots in the exploration of domestic and foreign unconventional natural gas. Since the end of 20th century, our country and foreign countries have made important progress in the exploration and development of coalbed gas. United States and Australia are the countries that have earliest launched out large-scale commercial development to coalbed gas in the world [71].

Early in the late period of 1970s and early period of 1980s, United States has become conscious that coalbed gas can be adopted as a kind of energy. In 1982, United States has launched out commercial development to the coalbed gas for the first time in Blackwarrior Basin of Alabama. A while later, United States main production base of coalbed gas San Juan Basin has started development. Then it has launched out coalbed gas development to Powder River Basin, Alabaqia Basin, Raton Basin, Uinta Basin, Pisens Basin and Green River Basin successively. In 2001, United States coalbed gas has realized a production volume of 7% of annual natural gas in the whole country. Since 1996, Australia has exploited two coalbed gas fields in Queensland and connected the produced coalbed gas with gas pipeline network.

Under the encouragement of production of United States coalbed gas, the people have aroused great interest in the development of coalbed gas. At present, various countries and regions have launched out early exploration or leading test, development and research into the coalbed gas apart from United States and Australia such as Russia, China, Canada, India, Germany, Poland, United Kingdom, Spain, France, Hungary, Dutch, Czech Republic, New Zealand, South Africa and Zimbabwe. Some of them are developing towards the commercialization [71].

As the second largest coal resources country in the world, China has realized a total coal resources volume of 5.57×10^{12} t (shallow coal resources with a bury depth of 2,000m) with a reservation volume of 1.02×10^{12} t and predicted reservation volume of 4.55×10^{12} t [72]. The enormous coal resources have constituted the important basis for the development of China's coalbed gas. According to the new round of national evaluation result of oil and gas resources, China has realized a total volume of coalbed gas resources of 37×10^{12} m^3 [15]—it is a little smaller than the normal natural gas resources in the land. But compared with the world technologically advanced countries, China's exploration and development degree of coalbed gas still has a large gap.

Through the exploitation, test and research of huge amount of coalbed gas in recent years, China has realized rapid technological development; all the exploitation normal technologies have been prepared and can realize implementation in matching. At present stage, it is importing and testing the development technology of pinnate horizontal well and exploitation and production increase technology of filled nitrogen and CO_2. It has successively conducted the exploration and exploitation test in Hebei Dacheng, Shanxi Liucheng and Jincheng, Liaoning Fuxin and Tiefa and Shaanxi Hancheng and attained rather good effect; besides, it has launched out test in Shanxi Jincheng, Liaoning Tiefa and Fuxin and drilled more than 1,000 production wells with an annual production volume of about $1 \times 10^8 m^3$ coalbed gas.

2.4.2 Oil sand

The oil sand is the general term of mixture formed by the crust surface clastic matters or rocks and contained viscous petroleum. Since various countries are conducting the resources statistics, they have certain differences within the scope of oil sand. The article has adopted the international widespread distribution plan. "Oil sand" mentioned here shall include the natural asphalt and overweight oil mineral. The following countries are rich in oil mineral resources: Canada, Former Soviet Union, Venezuela and United States [71]. Canada is the country with richest oil sand mineral resources which takes up about 77% of the global volume. Albert Basin is the main distribution area and includes eight oil fields such as Athabasca, Cold Lake and Peace River; it has a total geological reservation volume of about $(2,680-4,000) \times 10^8$t. Former Soviet Union has a oil sand mineral reservation volume of about $1,200 \times 10^8$t which takes up about 19% of the global volume; Venezuela has a oil sand mineral geological reservation volume of about $(490-930) \times 10^8$t; United States has a oil sand mineral geological reservation volume of about $(90-160) \times 10^8$t. At present, the world oil sand minerals have mainly been exploited through open air exploitation method; the key technology lies in the oil sand separation technology in the refinery. Venezuela has adopted the advanced oil pump and horizontal well technologies as well as improved thinner in the heavy oil development which has greatly upgraded the yield rate of heavy oil. United States has adopted the steam injection method in the heavy oil exploitation.

China is also one of the countries with rich oil sand mineral resources and its resources volume ranks fifth in the world. At present, it has discovered the oil sand mineral resources in numerous basins such as Bohai Bay Basin, Turpan-Kumul Basin and Tarim Basin. According to the new round of national evaluation result of oil and gas resources, China has realized a tar sand oil geological resources volume of 6 billion t and exploitable resources volume of 2.3 billion t [71]. Compared with Canada, Venezuela and United States, China has developed and utilized the tar sand oil at a later time; however, it has enjoyed rapid development in the heavy oil heat exploitation. At present, it has formed a rather mature matching technology of thick oil heat exploitation project and the development level has constantly upgraded; it has attained great achievements in various types of oil reservation, especially thick oil reservation with deep layer, numerous oil layers and inadequate homogeneity as well as steam injecting development. In 2005, China has realized an annual production volume of 180 million t of petroleum: the heavy oil has realized an annual production volume of 23.86 million t which takes up 13.2% of the national production volume of crude oil and about 1% of world petroleum volume. With the development and constant consumption of China's normal oil and gas resources, it will enjoy enormous development potential for the oil sand resources in the future.

2.4.3 Oil shale

Oil shale is solid mining. It is high ash flammable organic rock with oil content ranging from 4% to 20%. Liquid oil can be acquired through extraction. Oil shale has been found in many countries. United States, Brazil,

Estonia, China, Russia, Australia, Canada, Morocco, Israel and Jordan are countries[71] abundant in oil shale mining. Due to the lack of comprehensive and deep research, the evaluating results of various institutions and experts to the shale oil quantity are much different with each other, wherein the relative conservative predicted result of World Energy Council (WEC) [73] is from 2.48 trillion to 6.23 trillion barrels (from 339.726 billion to 853.425 billion t). However, due to technical and economical reason, only a few countries have made commercial exploration to the oil shale. Of all developed oil shale, 69% is for power generation, 25% is for extracting shale oil and 6% is for chemical industry and others.

Israel leads the world in oil shale exploration. In Israel, oil shale has been used as the fuel for power plant. However, since 1990s, Israel has developed a new technology of extracting oil from oil shale. That is to mix up the oil shale and regular refining residual for catalytic treatment. Due to the breakthrough in new technology, the extraction of liquid oil from oil shale by Israel has dropped from 50 US dollars/barrel to less than 20 US dollars/barrel, which has greatly increased the benefit of oil shale exploration. This technology reduces the imported crude oil of Israel by about 1/3.

China is one of the countries that are early in using oil shale. The oil shale was used for refining before the establishment of People's Republic of China and at the early stage of foundation. Two bases of Fushun and Maoming were built with annual output of tens of thousands tons and gradually shrunk and stagnated since some large oil and gas fields were found. Although China is abundant in the oil shale, the grade is low and the oil content varies much. According to the statistics of State Information Center in 2002, among the 63 regions in which the oil shale has been explored, the oil content of the oil shale of 49 regions ranges from 3.5% to 5%; that of 10 regions is larger than 10% with the highest value of 26.7%. The oil shale resource which has been found out is 33 billion t with technical exploitable coefficient of 56.4%. The technical exploitable oil shale resource found out is 18.6 billion t. The new oil and gas resource investigation result in China shows that the geological resource volume of shale oil converted by oil shale is 47.6 billion t and recyclable shale oil is 12 billion t [15]. Therefore, China has a certain potential in oil shale exploration in future.

2.4.4 Tight sand gas

Tight sand gas refers to natural gas within the sand reservoir which has low porosity (<12%), low permeability (<$0.1 \times 10^{-3} \mu m^2$), low gas saturation (< 60%), high water saturation and slow flow of natural gas in reservoir. Some scholars have raised some special tight sand gases such as "deep-basin gas", "basin center gas", "source-contacting gas" according to the unique characteristics of tight sand gases.

The tight sand gas almost exists in all oil and gas regions in the world. San Juan Basin in America was first found in 1927. The large tight sand gas field—

Elm Worth was found in Northern Depression in the west of Albert Basin in Canada in 1976. The tight sand gas reserve is huge. According to estimation, the current exploitable tight sand gas ranges from 10.5 trillion to 24 trillion m^3, the largest of all unconventional gas resources. According to statistics, there are 70 basins in which tight sand gas has been found or speculated to exist, mainly in North America, Europe and Asia-Pacific region. At present, the large tight sand gas deposit developed oversea is mainly deep-basin gas, mainly in Western Canada and Western America [74].

In China, the tight sand gas is widely distributed in Szechwan Basin, Ordos Basin, Turpan-Kumul Basin, Songliao Basin, Southern Dzungarian Basin, Southwest of Tarim Basin, Chuxiong Basin and East China Sea basin [75]. Since Zhongba in Western Sichuan was found in 1971, the research of tight sand gas has been made gradually in a systematic way. In 2008, the output of natural gas with low permeability including tight sand gas in China was 32 billion m^3, accounting for 42.1% [76] of the total output of natural gas of that year. The effective exploration and development of tight sand gas will be of great significance in ensuring the supply of natural gas in China.

2.4.5 Shale gas

Shale gas is a kind of natural gas which mainly exists in dark mud shale and silty mudstone stratum in forms of adsorption, dissociation or dissolution. These stones are generally tight, interbed or interlayer, thus facilitating the coexistence of dissociated natural gas and absorbed natural gas, with the important characteristics of introduction of absorption mechanism and intrinsic essence of nearby (local) concentration. The shale gas is continuously formed biochemical gas or thermal gas or the mixture of both. With universal stratum saturation, unknown concentration mechanism, various lithologic sealing and relative short transportation distance, the shale gas can dissociate in natural crack and pore, absorb on kerogen or surface of clay particle and even dissolve in kerogen and bituminous matter.

The exploration and development of shale gas was started in United States. The first industrial shale gas well was drilled in 1921 in United States. Since then, many gas reservoirs have been found successively in United States. In 1970s, United States government invested a lot in geological and geochemical research of shale gas and made great breakthrough in shale gas absorption mechanism research, thus greatly improving the output of shale gas of United States by 7 times from 1979 to 1999 [77]. The shale gas output of United States in 2005 was about 600 billion ft^3 (16.98 billion m^3) , accounting for 8% of total natural gas output of United States that year. The output of Newark East Shale Gas Field—the second largest gas field with Mississippian Barnett shale as the reservoir in Fort Worth Basin approached 14.16 billion m$^{3[78]}$ in 2005. So far, United States has more than 39,500 shale gas wells and shale gas has become one of the three unconventional natural gases (tight sand gas, coalbed gas and shale gas) which are put into industrial exploration and development in United

States.

The huge success of shale gas exploration in United States greatly stimulates the enthusiasm of different countries all over the world in searching natural gas resource in shale sequence. In recent years, some scholars have started to pay attention to the shale gas resource of China. However, the research hasn't been made systematically.

In Sichuan Basin, Ordos Basin, Bohai Bay Basin, Jianghan Basin, Turpan-Kumul Basin, Tarim Basin, Dzungarian Basin and other basins containing oil and gas and their circumference, the shallow dark (mud) shale develops in large area with high organic carbon content, thus having the favorable geologic conditions for shale gas reservoir. Meanwhile, the prospect of shale gas exploration in Cambrian system, Silurian system, Permian system and other ancient stratigraphic distribution areas in Southern China cannot be neglected.

2.4.6 Water-soluble gas

Water-soluble gas is a kind of unconventional natural gas with methane as main content[80]. The earth is abundant in water-soluble gas resource, which was about $400 \times 10^{12} m^3$ in former Soviet Union, $6.17 \times 10^{12} m^3$ (original resource) only in Texas and Louisiana of United States, and $(0.739 \quad 0.887) \times 10^{12} m^3$ in Japan. Besides, Hungary, Philippines, Nepal, Iran and Italy are also rich in water-soluble gas resource. The total amount of water-soluble gas resource in the world ranges between $n \times 10^{16}$ and $n \times 10^{18} m^3$. The daily gas productions of each well in different countries and regions are quite different, ranging from 100 to 40,000 m^3. So far, Japan is the country with most water-soluble gas exploration. The exploration amount reached $5.45 \times 10^8 m^3$ early in 1977 and exceeded $1.3 \times 10^{10} m^3$ in accumulation in 1978[81].

China has wide water-soluble gas distribution. Sichuan Basin (Weiyuan Gas Field), Songliao Basin, Bohai Bay Basin, Qiongdongnan Basin, Ordos Basin and Turpan-Kumul Basin have good prospect for water-soluble gas exploration. The water-soluble gas amount in China is estimated to be ranging from 12 trillion to 65 trillion m^3. The research and development of water-soluble gas in China lags behind. So far, the research of water-soluble gas has been carried out in Tarim Basin, Tianhe Gas Field and gas field in Middle Ordos.

2.4.7 Natural gas hydrate

Natural gas hydrate is a clathrate compound formed by methane and water under a certain temperature and pressure in permafrost region and deep ocean. The technologies for exploration of natural gas hydrate are still in the experimental stage (experiments are carried out in United States, Canada, Japan and Russia), mainly including depressurization, thermal excitation and compound breaking down. The coastal waters and Tibetan Plateau of China has the condition for forming natural gas hydrate.

Since 1990s, as more and more natural gas hydrate mines are found in the world and people know more and more about this energy mine, relevant researches have been carried out rapidly all over the world. The United States, Russia, Japan, Canada, United Kingdom, Norway, India and Pakistan have set up researches and made important discovery. On June 5th 2007, Ministry of Land and Resources released that China found natural gas hydrate in Shenhu Area of Northern South China Sea and got the sample. According to investigation, there is evidence for existence of natural gas hydrate in northern slope of the South China Sea, Nansha Trough Area, slope of East China Sea, Tibetan Plateau and permafrost in Northeast China. Besides, according to the latest news released by Ministry of Land and Resources on Sep. 25th 2009, natural gas hydrate sample has been found for many times in permafrost regions, Muli Town, Tianjun County, Qinghai Province in Nov. 2008 and Jun. 2009. However, it is still need to assess China's potential in natural gas hydrate.

In the coming years, the key scientific and technological problems on research of natural gas hydrate in China are as follows:

1）Define the existence, distribution and prospective amount of natural gas hydrate in China sea area and Tibetan Plateau.

2）The geological conditions and assessing method of natural gas hydrates of various types.

3）Exploration technology and method of natural gas hydrates.

2.4.8 Igneous reservoir

It has been about 100 years of igneous reservoir exploration since the igneous reservoir was found in Mexico Paleogene in 1907. A lot of igneous reservoirs have been found on the earth. This can be divided into four stages: "a forbidden zone people tried to avoid, confusion caused by discovery by chance, wandering of tentative exploration and objective actively searched for".

So far, it has been proved through exploration that the igneous reservoir has the characteristics of high yield, stable yield and a certain reserve scale and is the new field and direction for future oil and gas exploration. For example, the open-flow capacity of Xushen No. 1 Well and Shenshen No. 2 Well of Songliao Basin reaches 1 million m^3 and Shengshen No. 2 Well has stable yield for 7 consecutive years; the open-flow capacity of Changshen No.1 Well of Changling Oil Field reaches 1 million m^3; the single well yield of 1808 well of Basalt Oil Field, carboniferous, Karamay is 53.2t/d, the single well yield of 1809 well is 30.6t/d and the reserve volume is 100.41 million t; the single well yield of No. 1 well of Zhougongshan Gas Field in Sichuan Basin is 25.6×10^4m^3/d. All of the above shows the great potential of igneous reservoir. Meanwhile, it also shows that we should consider igneous reservoir in dialectically.

The igneous reservoir in China includes eruptive rock, irruptive rock and volcanoclastic rock, from basic to acid. With huge time span and depth from several hundred meters to 4,000m, characterized in late stage. Non-vibration geophysical techniques such as weight, magnet and electricity play an active role in exploration of igneous reservoir. However, the relationship between magmatic activity and hydrocarbon generation needs further research. We shall strengthen the relationship between basin structure and hydrocarbon generation on the basis of previous research and make research on the following scientific problems: the restriction to oil and gas by magmatic-volcanic activity, the restructure of igneous reservoir, the interaction of magmatic, liquid and solid, the formation mode of igneous reservoir and the geophysical technique for prediction of volcanic rock distribution.

What's more, we have found in abiogenic natural gas exploration that the effective identification of abiogenic natural gas, the formation mechanism of abiogenic natural gas and the environmental conditions, main control factors and rules for reservoir formation are important problems to be tackled. The defining of these problems will help us to deepen the understanding of formation rule and distribution of unconventional natural gas of China and promote the exploration and development of this kind of natural gas.

2.5 Oil and gas development

So far, main oil fields of China have entered a high water-cut stage and super high water-cut stage. The replaced resource is mainly low-grade special oil and gas reservoir. However, we don't have enough techniques and methods for further increasing recovery ratio in the high water-cut oil field which is in the leading position of crude output and has high dispersed surplus oil. We have a long distance away from foreign countries in stimulation and modification technology for low permeability oil and gas reservoirs and marine carbonate formational gas reservoirs. Generally speaking, main technologies restricting the stabilization and increase of oil and gas output of China are as follows:

2.5.1 Methods and technologies for increasing water drive recovery factor

The surplus oil of mature oil fields in Eastern China is highly dispersed. The oil-water well casing damage is high. There are a lot of reservoir well damage, with less perfect strata sequence and well pattern. The water drive control degree shall be further increased. The technology for refined reservoir description and the surplus oil prediction is imperfect and a new subsurface reservoir recognizing system is needed. After long-term water flood development, a preferential channel has formed and caused the low efficient and inefficient circling of flood water.

2.5.2 Development of high-efficiency chemicals for EOR

The research of polymer injection recovery ratio increase of foreign countries was popular from 1970s to 1980s. Since 1986, due to the increase of environmental awareness, high cost of polymer injection and drop of oil price, the proportion of polymer injection for oil production in foreign countries has dropped and the relevant researches have gradually become less concerned.

However, in recent years, polymer injection oil drive technology has developed rapidly in China and now leads the world. It has been popularized and applied in large scale and become an important means of increasing the recovery ratio by large extent in China. However, since 60% surplus oil is widely distributed in reservoir after water drive of polymer injection and the distribution in horizontal direction and vertical direction is quite complex, we are faced with the bottle-neck problems such as the quantitative explanation of surplus oil distribution after polymer injection and complex drive techniques for strong base, weak base and non-alkali surface active agent system of different reservoirs.

2.5.3 High-efficiency reformation techniques for low permeability and ultra-low permeability oil and gas field reservoir

The proportion of low permeability reserve in explored reserve ranged from 40% to 50% between 1990 and 1995. It has been kept between 60% and 65% since 1995. In 2006, it rose to 67%. Low permeability reserve is still the main part of newly explored reserve. The low permeability reserve accounts for 2/3 of undeveloped reserves. The main body of productivity construction of the new area is low permeability reservoir. For low permeability and ultra-low permeability oil and gas field and fracture-cavity carbonate oil and gas field, the effective identification and prediction methods for cracks relative to high permeability reservoir development region are lacked. The high-efficiency reformation techniques of reservoir are imperfect and the further research on accessories is needed.

2.5.4 High-efficiency development techniques for fracture-pore carbonate oil field

The fracture-pore carbonate oil field has complex geological conditions. The cracks and solution cave are developed. The reservoir is heterogeneous. We shall pay attention to the characteristics and distribution law of fracture-pore reservoir, reservoir identification and prediction and high-efficiency development matching techniques, mainly including fracture-pore reservoir identification and prediction, complex fissure medium value simulation techniques and software, high-efficiency acid reformation techniques of carbonate reservoir.

2.5.5 Engineering technology of deep water oil and gas exploration and development of China

China is far away from other countries in deep ocean engineering technology, which can be embodied in that the deepest oil field developed by China in cooperation is 332m, far behind the world record 2,400m. The biggest capacity of drilling equipment in China is only 503m while the largest in the world is above 3,000m. So far, the oversea deep water operation capacity of offshore engineering is 3000m and the largest crane capacity is 14,000t while the operation capacity of offshore engineering equipment in China is within depth of 150m and the largest offshore crane capacity is 3,800t. The huge discrepancy in deep water equipment and technology has become the bottle-neck of the exploration and development of deep water oil and gas resource of China.

2.5.6 Industrial production technology of shale oil

At present, the cognitive level of mineralizing and enrichment law of shale oil is relatively low. There is no systematic and targeted technological specification, methodological system and parameter standard for the appraisal of shale oil resource. Effective technology is lacked for the underground exploration of shale oil resource, such as the in-situ retorting of the shale oil resource buried deep in the big basin. Therefore, the technologies of shale oil potential appraisal and advantageous target optimizing, the in-situ retorting of shale oil and comprehensive application of shale oil are urgently needed in shale oil exploration and development.

2.6 Instruments and equipments

2.6.1 Large high-precision digital seismograph

With regard to the technical difficulty of geophysical exploration of petroleum, digital geophone and digital seismograph with intellectual property rights shall be developed to help us master the technologies of manufacture of the high-precision, high-resolution and high-signal to clutter ratio seismic exploration equipment and data collection. Meanwhile, it makes the main technological index of the seismic exploration instruments of China reach the international advanced level; seismic exploration technology series and industrial structure which solves the complex geological structure shall be formed to meet the requirement of the development of oil and gas geophysical exploration theory and technology, increase the comprehensive technical capacity for R&D and application of geophysical equipments and enhance the core competency of geophysical equipments of China.

With adoption of industrial standardized design and combination of wire and wireless, the large-scale high precision digital seismograph—all digital seismic data transmission and recording system which is based on independent intellectual property rights is suitable for different seismic exploration and geological conditions.

The digital geophone based on the micro engine machanical system (MEMS) has better amplitude response and frequency response than regular geophone and larger dynamic range; high density single-receiver is good for processing seismic outdoor data, clearly describing underground structure, lithology and oil-gas possibility and lowering exploration and development risk.

Due to the complex geological conditions and the existence of deep oil and gas reserve, the large high-precision digital seismic exploration system must reach 1 million and meanwhile realize wireless communication of the same scale, thus ensuring the seismic stacking fold is better than present 180 thousand to 300 thousand and the high-resolution detection of deep oil and gas and complex geological body can be realized. So far, the chip of digital geophone based on the micro engine machanical system (MEMS) has reflected the restriction of the intellectual property rights on the product from the price.

2.6.2 Deep marine streamer seismic collecting vessel and retractile system

Streamer seismic exploration is the main method adopted for deep marine oil and gas exploration. The market share of deep marine streamer seismic collecting market is above 40% of seismic exploration market all over the world. The development of deep marine streamer seismic collecting vessel and matching seismic collecting equipments has vital significance for enhancing the comprehensive ability of offshore seismic exploration of China and improving the competitiveness in international offshore seismic exploration market.

Deep Marine Streamer Seismic Collection

It is a method of offshore seismic exploration by using of exploring ship, characterized in excitation and receiving in water and uniform excitation and receiving conditions; continuous observation without stopping ship can be conducted. The non-explosive source is mainly adopted and piezoelectric seismometer is often adopted for receiving. During work, the seismometer and the cable shall be dragged and dropped in the sea water of a certain depth.

Since the streamer seismic collecting equipment is highly professional and technology intensive, at present, its manufacture is monopolized by some western professional companies, such as equipment manufacturing company of French CGG and I/O of the United States. There is no relevant equipment design and manufacture capacity in China.

Besides, the software solidification, comprehensive geophysics and drilling technology of deep water oil and gas exploration instruments will become the direction of development in future.

2.6.3 Deep water drilling platform

With the rapid development of world marine oil and gas to the deep sea and super deep see, the demand for deep water drilling platform becomes larger and larger in the world. The deep water oil and gas exploration of China is still in the initial stage. The shortage of deep water drilling platform strictly restricts the exploration and development of deep water oil and gas in China.

The working depth of the first semi-submersible drilling platform designed and constructed by China independently is only 200m; the maximal working depth of drilling platform brought in from oversea is only 610m while that of the most advanced drilling platform in the world is above 3,000m.

The largest working depth of the independently developed deep water drilling platform is above 3,000m, which cannot only meet the need of China offshore deep water oil and gas exploration, but also greatly increase the competitiveness of China in oil and gas exploration in international deep water field.

3 Prediction of Development Objectives in Scientific Field on Oil and Gas Resource in the Future

3.1 Overall description to scientific roadmap in oil and gas resource field to 2050

Researches show that China has rich oil and gas resource. However, there is still a certain distance to oversea advanced level in overall exploration effect. We are faced with of the problems of "low reserves abundance, deep occurrence and hidden target". The future oil and gas exploration shall develop from structural trap to non-structural lithologic and stratigraphic trap, from continental strata to marine strata, from shallow to deep and from onshore exploration to offshore exploration. We need to make full use of the achievement of global science and technology to continuously solve the theoretical and technical problems faced by natural gas exploration and form a theoretical system and core technology through independent innovation, thus ensuring the stable development of oil industry of China. We should adhere to the energy strategic principle in the next few decades of "giving priority to the exploration and development of domestic oil and gas resource, moderately importing oil and gas according to market rule, building energy saving society and developing renewable clean energy", which is also the guarantee of maintaining independence in terms of politics and diplomacy and safeguard national interests and also the outline for guiding the scientific and technological development of oil and gas.

The overall route map of scientific and technological development of oil and gas to 2050 suitable for conditions of China has been made according to the research of characteristics and targets of sustainable oil and gas system construction of China to 2050 (Table 3-1) with combination of special geological background and the forefront of oil and gas scientific and technological development of the world. According to time period, the scientific and technological development goal of oil and gas in China can be divided into shortterm goal, middle-long term goal and long term goal. (Fig. 3-1).

Table 3-1　Characteristics and Targets of Sustainable Oil and gas System Construction of China to 2050

Item	around 2025	around 2035	around 2050
Oil and gas exploration	exploration ratio of crude oil: 50% exploration ratio of natural gas: 30% exploration depth: 8000m	exploration ratio of crude oil: 60% exploration ratio of natural gas: 50% exploration depth: 10,000m	exploration ratio of crude oil: 70% exploration ratio of natural gas: 60% exploration depth: 12,000m
Oil and gas application	recovery ratio of crude oil: 40% Unconventional oil and gas replacement ratio: 10%	recovery ratio of crude oil: 50% Unconventional oil and gas replacement ratio: 20%	recovery ratio of crude oil: 60% Unconventional oil and gas replacement ratio: 30%-40%

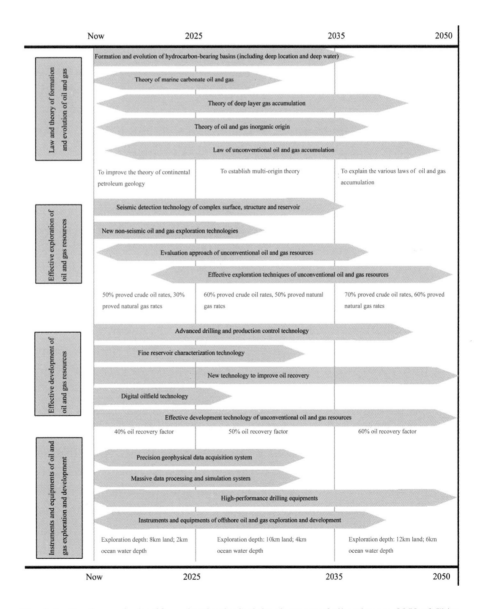

Fig. 3-1　Roadmap of scientific and technological development of oil and gas to 2050 of China

3.1.1 Short-term Targets (around 2025)

The scientific and technological level in overall exploration and development of China shall be near to the advanced level in the world through a series of scientific and technological progress in oil and gas resource. The exploration ratio of crude resource shall reach 50%, the natural gas 30% and the crude recovery ratio 40% (Fig. 3-1).

- Deepen the fundamental geological research of basin to discloses the restriction of temporal and spatial evolution of global structure on development characteristics of basins of various types, the relationship between the deep fault structure and oil and gas reservoir, the lithologic-stratigraphic oil and gas formation mechanism and the geological law of hydrocarbon generation kinetics mechanism of the basin and to perfect the theoretical system of basin formation and evolution and the oil and gas reservoir.

- Set up marine carbonate rocks oil and gas reservoir formation system initially and make research on the distribution laws and formation mechanisms of Pre-Cenozoic marine carbonate rocks oil and gas which has experienced later tectogenesis damage for many times to improve the methods and technologies for exploration and development of marine carbonate rocks oil and gas and improve the effect of exploration and development of carbonate bed series.

- Make quantitative research on the deformation and reservoir formation of oil and gas containing basin with complex evolutionary history through setting up a basin formation dynamics model. Set up technologies for description and prediction suitable for reservoirs of various types (including marine carbonate sequence and low permeable reservoir) in China.

- Make further breakthrough in sequence stratigraphy. Further recognize the formation mechanism and distribution law of lithostratigraphic reservoir and hidden reservoir. Further improve the stratigraphic precision interpretation from shallow exploration to deep exploration and from clastic strata to carbonate rocks and volcanic rocks.

- Set up geophysics detective technology and equipments for complex ground surface, complex structure and complex reservoir.

- Develop high-resolution seismic measurement and interpretation technology to solve the formation mechanism of low permeable, fissure and unconventional reservoir and the detection precision of deep and high quality storage body spatial distribution.

- Work on the surplus oil distribution mechanism and develop biological recovery technology. Increase the recovery ratio by large extent with the secondary and tertiary recovery technology.

- Set up exploration and development technology of offshore deep water-super deep water basin with intellectual propriety rights.

- Verify the existence of methane hydrate deposit in ocean area and permafrost regions of Tibetan Plateau of China. Work on the mechanism and environment for the formation and preservation and make trial industrial exploitation.
- Form the high-efficiency exploitation technology of asphalite, oil sand, oil shale and other unconventional oil and gas mining.

3.1.2 Medium-long term targets (around 2035)

The scientific and technological level in overall exploration and development of China shall be near to the advanced level in the world through a series of scientific and technological progress in oil and gas resource. The exploration ratio of crude resource shall reach 60%, the natural gas 50% and the crude recovery ratio 50% (Fig. 3-1).

- Make big progress in research on distribution law of oil and gas containing basin of China from the angle of global structural evolution pattern to basically master the formation type and distribution law of the oil and gas containing basin of China.
- Have deep reorganization on relationship between deep water-super deep water basin deep structure and stratum developing characteristics and form a complete set of oil generation theory, reservoir formation theory and petroleum system theory.
- Master the distribution law and formation mechanism of hydrate in ocean area of China and set up the safety and environmental protection technology of exploration and development of methane hydrate deposit.
- Conduct deep water to ultra-deep water oil and gas exploration independently.
- Set up a comparatively complete deep oil and gas reservoir forming theory initially and set up technology of deep layer to ultra-deep layer oil and gas drilling and development.

3.1.3 Long-term targets (around 2050)

The scientific and technological level in overall exploration and development of China shall be near to the advanced level in the world through a series of scientific and technological progress in oil and gas resource. The exploration ratio of crude resource shall reach 70%, the natural gas 60% and the crude recovery ratio 60% (Fig. 3-1).

- Carry out oil and gas exploration and development of deep water to ultra-deep water basin in an all-round way.
- Methane hydrate deposit enters the stage of large scale development.
- Realize the deep-layer and ultra-deep-layer oil and gas drilling and development to greatly increase the exploration effect and recovery efficiency of deep layer oil and gas by large extent.
- Participate in the oil and gas exploration and development in Arctic Circle and other public areas of the world.

3.2 Theories of formation and evolution of oil and gas

3.2.1 Roadmap of technological development

(1) Overall targets to 2050

Efforts shall be made to solve the problems of the formation and evolution of oil and gas containing basin, the reservoir formation of marine carbonate rocks oil and gas, the deep basin reservoir formation, the inorganic petroleum origin and the unconventional oil and gas accumulation through the research with the onshore and offshore of China as key area and also combining the oversea, deep ocean, open ocean and polar region and other public regions, with structure and basin formation and evolution, subsurface fluids and oil migration and aggregation, great geological event and oil and gas reservoir, the coexistence of oil and gas and other energy mineral resources, oil and gas distribution controlled by different structural units of global-region-basin, spatial and temporal distribution and effective identification mark as the main content and by means of basin analysis, structure analysis, sequence contrast, isotopic analysis, oil/gas source tracing analysis, experiment analysis and simulation and mathematical modeling, with the aim of setting up a sound oil and gas geological theory system to provide theoretical guidance to oil and gas exploration. (Fig. 3-2).

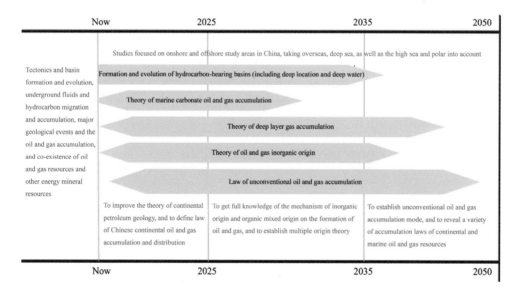

Fig. 3-2   Scientific and Technological Development Roadmap of Oil and gas Formation and Evolution Law and Theory

Objectives at the stages

To enrich and perfect the continental petroleum geological theory which has continental oil generation theory, source-controlled theory, complex oil and gas accumulation zone theory, oil-rich depression "full depression with oil" and structure-sequence composition as the core and define the law of continental oil and gas reservoir formation and distribution before 2025; to have enough recognition to the formation mechanism of inorganic origin and mixing origin of inorganic and organic oil and gas and set up perfect multi-genesis theory before 2035; to set up reservoir formation mode of unconventional oil and gas and discover the formation law of various onshore and offshore oil and gas resources.

3.2.2 Main scientific questions Involved

(1) Formation and evolution law of oil and gas containing basin

Basin is an important site for oil and gas reserve. Therefore, research on formation and evolution of oil and gas containing basin is of vital importance to exploration and development of oil and gas resource. Under the combined action of three dynamical systems of Ancient Asian Ocean Tectonic Domain, Ancient Tethys Ocean Tectonic Domain and Circum-Pacific Ocean Tectonic Domain, cratons of different sizes and ages together with the folded orogenic belt of different ages forms various basins and basin fills of China. The variety causes the complexity of formation and evolution of oil and gas containing basin of China. In order to solve this problem, we must first tackle the following issues: ①formation and evolution of superimposed basin; ②set up structural models of basin in different periods in many ways; ③complex structure geological model of Midwest foreland thrust belt; ④recovery of basins of various prototypes; ⑤relationship between basins of different genetic types and oil and gas accumulation.

(2) Marine carbonates rocks oil and gas reservoir formation theory

Marine hydrocarbon theory has been raised for over a hundred years and a lot of oil and gas exploration of marine strata has been carried out oversea. However, different from oversea marine strata, most of the Pre-Cenozoic marine strata of China exist in the form of marine residual basin due to old age and much reformation of complex structure and have the prevalent characteristics of deep bury, complex fault system, various types, multi-source, loosely distributed and strong concealment, which leads to the difficulty in oil and gas exploration. Therefore, it has never been taken as a key area in China for long. Due to the improvement of technical measures, a plurality of large-middle sized oil and gas reservoirs have been found in marine carbonate strata in recent years and marine carbonate oil and gas reservoir has become a hot point of exploration again. However, due to the long term neglect, the theoretical research of marine carbonate oil and gas reservoir of China lags behind and the research on following issues shall be carried out: ①oil generation of marine carbonate; ②formation mechanism of buried hill, organic reef and other carbonate reservoirs; ③spatial distribution and interrelation of different carbonate reservoirs; ④dolomitic mechanism of carbonate reservoirs; ⑤ reservoir formation mould of marine carbonate.

Deep layer reservoir formation theory

With the continuous oil and gas exploration and development, it is difficult to find large-scale oil and gas reservoir in middle and shallow layer of basin. Therefore, people gradually turn to the deeper layer of the basin. The complexity of continental strata structure and sedimentary evolution leads to the complexity and specialty of occurrence condition and formation law of deep layer oil and gas. The scientific problems needing to be solved for research include:①formation and reformation of deep layer of basin;②evaluation of deep layer high-post mature hydrocarbon source rock;③prediction of formation and distribution of effective reservoirs in deep layer;④oil and gas reservoir formation mode of deep layer;⑤evaluation of advantageous exploration region of deep layer.

Oil and gas inorganic origin theory

Oil and gas inorganic origin is a complex theoretical issue which hasn't been solved completely. Although the organic origin theory is dominant in oil and gas geological exploration at present, however, with the development of scientific research and oil and gas exploration in recent years, some natural gas reservoirs with mantle source or mixed source characteristics (mainly CO_2 and little hydrocarbon gas) have been found successively, and thus inorganic origin becomes a hot point for oil and gas geologic research again. The scientific problems needing to be solved in future include:①generation mode of inorganic origin oil and gas;②accumulation mechanism of inorganic origin oil and gas;③relation between distribution and fault of inorganic origin oil and gas; ④identification criteria for inorganic origin oil and gas;⑤reservoir formation mode of inorganic origin oil and gas.

Law of unconventional oil and gas reservoir formation

Under the combined action of continuous consumption of regular oil and gas resource, rise of oil and gas price and the progress of exploration and development technology, unconventional oil and gas has gradually drawn people's attention. Some unconventional oils and gases have already played an important role in energy structure. However, people's knowledge to unconventional oil and gas reservoir formation law is very limited. The research of following scientific problems needs to be carried out:①unconventional oil and gas reservoir formation mechanism;②relationship between unconventional oil and gas reservoir and the structure and strata;③distribution law of unconventional oil and gas reservoir;④unconventional oil and gas reservoir formation mode.

3.3 Effective exploration of oil and gas resources

3.3.1 Roadmap of technological development

(1) Overall targets to 2050

Taking all geologic units suitable for occurrence of conventional and unconventional oil and gas resources as objects of study, under the guidance of research findings in respect of geological theory on oil and gas, comprehensively conduct the innovation and integrated application of such prospecting techniques as seismic prospecting, magnetic prospecting, electrical prospecting, gravity prospecting, geochemical prospecting, remote sensing prospecting, microbial prospecting, etc. Improve the technology for acquisition of real information at oil and gas reservoirs and deep reservoirs under complex ground surface conditions, improve the review technology on oil and gas resources including unconventional oil and gas resources and explore effective special techniques on prospecting of unconventional oil and gas resources, thus providing theoretical and theological support for accurately evaluation of our country and its participation for development of the oil and gas resources of the regions(Fig. 3-3).

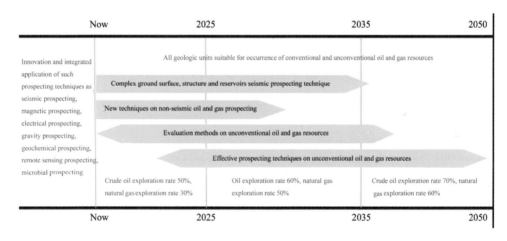

Fig. 3-3   Roadmap of Technological Development in Effective Prospecting of Oil and Gas Resources

(2) Objectives at the stages

Before 2025, through the improvement of the prospecting techniques and the data interpretation method, the crude oil exploration rate shall reach up to 50%, and natural gas exploration rate up to 30%; before 2035, the oil and gas prospecting techniques and technology composition methods shall be developed according to different types of oil and gas resources, so as to make the crude oil exploration rate reach up to 60%, and the natural gas exploration rate up to 50%; and before 2050, the integration of different prospecting techniques shall be strengthened while developing new prospecting techniques, so as to make our country's crude oil exploration rate reach up to 70% and natural gas exploration rate up to 60%.

3.3.2 Related major scientific and technological problems

(1) Complex ground surface, structure and reservoirs seismic prospecting technique

Our country is large in area, with complex ground surface conditions in some oil and gas exploration areas. It is also characterized by many types of surface configuration, such as loess plateau surface, hilly surface, desert surface and mudflat surface, etc. And in some regions, the conditions of both surface and underground are complex, such as the exploration area with high inclined formation at the front of the foreland basins[82]. The complex surface condition brings about difficulties for the accurate collection of physical data of the earth. Meanwhile, the complex geologic environment and deposition history of the most basins with oil and gas in our country also bring about difficulties for the interpretation and forecasting of physical imaging of the complex underground structure and the interpretation of complex oil and gas reservoirs. The major scientific and technological problems needed to be solved in respect of the physical prospecting techniques for complex surface, structure and reservoirs are: ① technology on collection of seismic data under complex surface condition; ② seismic imaging technology on underground complex structure, especially the imaging technology on high inclined formation; ③ technology on interpretation of seismic data under complex structure; ④ thin interbed seismic prospecting techniques; and ⑤ seismic forecasting technology on complex oil and gas reservoirs.

(2) New techniques on non-seismic oil and gas prospecting

There is no denying that seismic technology has played a leading role in the oil and gas exploration. However, some non-seismic prospecting technology have played an important role in the oil and gas exploration, especially in the regions where the seismic prospecting technology are difficult to be conducted. Compared with expensive seismic prospecting, the non-seismic prospecting costs less in capital. At present, as the improvement of the observation accuracy of the devices and instruments and the progress of the treatment of interpretation techniques, the non-seismic prospecting technology has further enlarged and deepened the application in the different stages of exploration and development, besides the continual application in early basin exploration and evaluation. The application of the non-seismic techniques has been changed from original area exploration to the zone exploration and target exploration (the key points are allocated at high inclined formation, areas with lava and deep hilly zones); extended from original structural configuration study to the oil and gas prediction evaluation (including oil and gas prospect evaluation, layout of reservoirs and prediction on types of oil and gas contained); extended from original oil and gas exploration to oilfield development and monitoring (applied on identification of affecting range of water, profile control effect evaluation and prediction on layout of rest oil and gas, etc.)[83]. The major scientific and technological problems needed to be solved in respect of new techniques on non-seismic oil and gas prospecting are:①new techniques on chemical prospecting for oil and gas;②microbe detection technique on oil and gas;③hyper-spectral remote sensing techniques on prospecting of oil and gas;④high-precision three-dimensional non-seismic oil and gas prospecting techniques; and⑤techniques on prediction and description of non-seismic reservoirs.

(3) Evaluation methods on unconventional oil and gas resources

The unconventional oil and gas resources are rich in the world. However, it prevalently lacks of the knowledge on both emplacement mechanism and occurrence of the unconventional oil and gas resources. What's more, the evaluations on the unconventional oil and gas resources between different personnel and between different authorities are different to a great extent. How to scientifically conduct reasonable evaluation on the unconventional oil and gas resources is a key factor to restrict human to develop and apply the unconventional oil and gas resources. The major scientific and technological problems needed to be solved in respect of unconventional oil and gas resources are:①the determination of the standard of recoverable reserves of the unconventional oil and gas resource economy;②evaluation mode on unconventional oil and gas resource reserves; and③accurate evaluation on reserves of all types of unconventional oil and gas resources.

(4) Effective prospecting techniques on unconventional oil and gas resources

Due to the different forms of emplacement mechanism and occurrence, the prospecting of unconventional oil and gas resources differs in great extent from that of conventional oil and gas resources. And it has some special features. For example, it often has BSR phenomenon in the seismic prospecting of marine gas hydrate, which can be one of the signs to indicate the existing of gas hydrate. In the exploration of lava oil and gas reservoirs, it often has high natural gamma values. The major scientific and technological problems needed to be solved in respect of effective prospecting techniques on unconventional oil and gas resources are:①techniques on seismic prospecting of unconventional oil and gas resources;②techniques on non-seismic prospecting of unconventional oil and gas resources;③techniques on interpretation of reservoirs of unconventional oil and gas resources; and④techniques on prediction of reservoirs of unconventional oil and gas resources.

3.4 Effective development of oil and gas resources

3.4.1 Roadmap of technology development

(1) Overall targets to 2050

Taking both conventional and unconventional oil and gas reservoirs that are under development and to be put into development as objects of study, under the guidance of research findings in respect of geological theory on oil and gas, comprehensively conduct innovation and integrated application of such production technique as well drilling, well pattern design, detailed description on oil reservoir, oil layer profile control, intensifying of oil production, etc. Form such special techniques and combination techniques on specific oil reservoir description techniques, high-water-cut oil reservoir development techniques, low-permeability oil reservoir development techniques, thick oil development techniques, etc. Explore effective techniques on development of unconventional oil and gas resources, thus providing theoretical and theological support for increase of the recovery ratio of our country and its participation for development of the oil and gas resources of the regions (Fig. 3-4) .

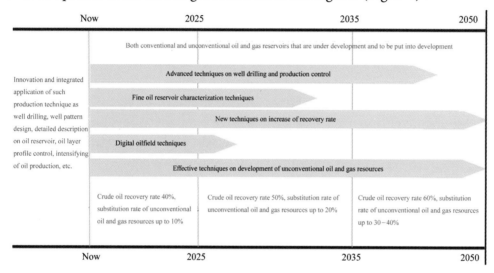

Fig. 3-4   Roadmap of Technological Development in Effective Development of Oil and Gas Resources

(2) Objectives at the stages

Before 2025, through the progress of drilling technique and the improved oil recovery (short for IOR), our country's crude oil recovery rate shall reach up to 40%, and substitution rate of unconventional oil and gas resources up to 10%; before 2035, through enhanced oil recovery (short for EOR) and the progress of techniques on precision management of oil-gas field, our country's crude oil recovery rate shall reach up to 50%, and substitution rate of unconventional oil and gas resources up to 20%; and before 2050, through the integration of different development techniques and the multi-path increase of recovery rate, our country's crude oil recovery rate shall reach up to 60%, and substitution rate of unconventional oil and gas resources up to 30–40%.

3.4.2 Related major scientific and technological problems

(1) Advanced drilling and production control technology

Oil and gas resources generally occur in comparatively deep layers underground and may be exploited by drilling. Pushed forward by related technology, the current drilling employs the form of straight-hole drilling, and is gradually developing into a state that straight-hole drilling and directional drilling coexist, so as to promote injection rate of oil-gas fields while reducing the unit cost of exploiting oil and gas and providing important support for exploiting marginal oil and gas resources. Meanwhile, advanced production control technology has made important contributions to promote the benefits from development of oil-gas fields. Crucial scientific and technological problems that need to be solved with respect to advanced drilling and production control technology in future are: ① drilling technology of horizontal wells and multi-branch wells; ② large displacement drilling technology; ③ small well hole drilling technology; ④ laser drilling technology; and ⑤ intelligent drilling technology.

(2) Detailed characterization technology of oil reservoirs

Oil reservoirs in the world are of great variety and complexity, most typical of which are the continental reservoirs whose non-uniform nature leads to non-uniformity of oil driven by water during the water injection process. When the reservoirs enter the stage with high water content, the more water-contented layers are, the more complicated relationship between oil and water underground and the bigger difference in the degree of inundation. Whether good effect of comprehensive adjustment can be achieved in development of oil fields depends to a large extent on the knowledge of oil reservoirs. Crucial scientific and technological problems that need to be solved with respect to detailed characterization technology of oil reservoirs in future are: ① dynamical monitoring technology of oil reservoirs; ② detailed simulation technology of oil reservoirs; ③ detailed description technology of oil reservoirs; and ④ predicting technology of remaining oil distribution.

(3) New technology to promote recovery ratio

The recovery ratio is a technological and economic index to measure the efficiency of oil exploitation. Oil exploitation technology can be classified into technologies for primary, secondary and third oil exploitations. Primary oil exploitation is carried out by making use of natural energy (i.e. pressure energy) in the oil layer; secondary oil exploitation is done by recovering pressure of the oil layer using certain processing technology, and the commonly used method is to exploit oil by injecting water; to promote recovery ratio belongs to third oil exploitation which includes enhanced oil recovery (EOR) developed in recent years. The new technology of increase recovery ratio is mainly used in the third oil exploitation. Crucial scientific and technological problems that need to be solved with respect to new technology for increasing recovery ratio in future include: ① improved flooding technology of intelligent polymer; ② alkali-surfactant-polymer (ASP) chemical flooding technology; ③ CO_2 capture & storage and flooding technology; ④ microbial enhanced oil recovery technology; ⑤ combined technology for effective exploitation of low permeability oil and gas fields.

(4) Digital oilfield technology

The application of such emerging and cut-edge scientific technologies as human-computer interactive workstation technology, colored image displaying technology, network technology, database technology, geographic information system and others has pushed forward the development of digital oil field technology and enhanced the analysis of general information of oil fields, and thus meets the need for deepening exploitation and development of oil fields. Besides, analysis of decision-making on operation and management of oil fields is also generally supported to further explore potential values in the process, creating a good information-supported environment for sustainable development of oil field enterprises. Crucial scientific and technological problems that need to be solved with respect to digital oil field technology in future include:①integration technology of multiple kinds of information; ②virtual reality technology; and③intelligent oil field technology.

(5) Effective development technology of unconventional oil-gas resources

The problem of how to effectively develop unconventional oil-gas resources has been checking people in making use of the resources on a large scale. Therefore, it is an objective that people have been striving for to realize the maximum development of unconventional oil-gas resources economically and technologically with environment-friendliness. Crucial scientific and technological problems that need to be solved with respect to effective development technology of unconventional oil-gas resources in future include: ①heavy oil well head and underground upgrading technology;②Vapex heavy oil absorbent extraction technology;③effective development technology of coalbed gas;④effective underground development technology of oil shale; and ⑤underground decomposition technology of natural gas hydrate.

3.5 Instruments and equipment for exploitation and development of oil and gas

3.5.1 Roadmap of scientific and technological development

(1) Overall targets to 2050

Take the various instruments and equipment for exploitation and development of oil and gas as the research objects; target at promoting the benefits of exploitation and development of oil-gas resources; aim to develop our manufacture of instruments and equipment for oil-gas exploitation and development to be of internationally advanced level by developing and producing such instruments and equipment with independent intellectual property rights as digital detector, digital seismograph, high performance equipment for detecting wells, advanced drilling equipment and advanced instruments and equipment for exploitation and development of oil and gas under the sea, so as to satisfy the demands of different environments under which oil and gas are exploited and developed, providing necessary support in instruments and equipment for oil-gas exploitation and development in China (Fig. 3-5).

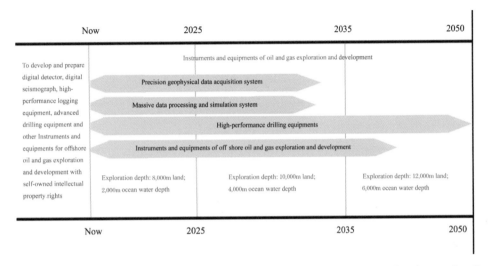

Fig. 3-5　Roadmap of scientific and technological development of instruments and equipment for oil-gas exploitation and development

(2) Objectives at the stages

By 2025, technological innovation in instruments and equipment may enable our exploitation goes to 8,000 meters deep into the land, and 2,000 meters water deep in the sea water; by 2035, technological innovation in operation of instruments and equipment and explanatory software may enable our exploitation furthers to 10,000 meters into the land and 4,000 meters water deep in sea water; by 2050, comprehensive and integrating innovation may enable our exploitation reach 12,000 meters into the land and 6,000 meters water deep in sea water.

3.5.2　Related major scientific and technological problems

(1) Highly precise global physical data collecting system

Focusing on the technological difficulty in global physical exploitation of oil and gas, we aim to develop and research instruments and equipment with independent property rights for collecting global physical data, master technology for manufacturing highly precise global physical instruments with high resolution and signal-to-noise ratio, and achieve global physical data collecting system for complex geological structures. Crucial scientific and technological problems that need to be solved with respect to highly precise global physical data collecting system include:①high-performance digital detector;②high-performance digital seismograph;③high-performance instruments for detecting wells; and④wireless transmission technology for global physical data.

(2) Mass data processing and simulating system

Global physical measures always generate mass data, and processing and simulation of these data are closely related to the exact analysis and interpretation of the measured data. With the fast development of modern computer and information technologies, technologies of data processing and simulating are also developing quickly. Crucial scientific and technological problems that need to be solved with respect to mass data processing and simulation system include:①storing technology of mass data;②analyzing software system of mass data;③simulation software system of mass data.

(3) High-performance drilling equipment

Drilling equipment is among the most important instruments and equipment in oil-gas field development and its improvement may greatly facilitate oil-gas field development. As the exploitation and development of oil and gas go forward, drilling technology tends to be dedicated to deeper exploitation, smaller well bore, faster speed and lower cost. Crucial scientific and technological problems that need to be solved with respect to high-performance drilling equipment include:①laser drilling equipment;②drilling equipment for deep wells and deepest wells;③drilling equipment for wells with small well bores;④drilling equipment for horizontal wells and multi-branch wells.

(4) Instruments and equipment for exploitation and development of oil and gas in sea

To push forward the exploitation and development of oil and gas in sea is a significant direction that future oil-gas industry will take. Compared with that on land, exploiting and developing oil and gas in sea need some special instruments and equipment, such as drilling platform, seismic data collecting boat and others. Crucial scientific and technological problems that need to be solved with respect to instruments and equipment for exploiting and developing oil and gas in sea include:①dragging and releasing of seismic data collecting boat that collects seismic data of deep sea;②drilling platform for exploitation and development in deep sea;③transmission system of oil and gas in deep sea.

4 Primary Scientific Research Orientations in Terms of Oil and Gas Resources by Chinese Academy of Sciences

"Based on exploration and development of domestic oil and gas resources; properly importing oil and gas in accordance with market rules; and making efforts to build energy-saving society and to positively develop renewable clean energy resources" is the strategic principles for oil and gas energy we must adhere to in the coming decades, and it's also the overall agenda to guide the development of oil and gas resources scientific technology. In the process of enhancing the science and technology of Chinese oil and gas resources exploration and development to achieving and maintaining foreign advanced level, Chinese Academy of Sciences should and is capable of making important contributions with its own characteristics and strengths. Analyzing from system, field configuration, research accumulation and other aspects of Chinese Academy of Sciences currently, it should focus on efforts in the following aspects to form a superior force in the next 15 years.

4.1 Tectonic evolution and marine oil and gas resources in China

The system collects and comprehensively summarizes basic geological research results worldwide, studies the establishment and process of the formation, drift, tectonics and transformation of Chinese continent in depth with the background of global tectonic evolution; it especially focuses on the restraints of Tethys to the marine oil and gas resources in China.

Preliminary study results show that the huge problems for exploration and development on deep burying, complex and various fault system, a variety types of oil resources, scattered distribution and strong concealment generally exist in oil and gas mineral resources in Pre-Cenozoic marine residual basin. From the innovative research achievements and exploration practices some time ago we can figure out that the oil and gas mineral resources at Pre-Cenozoic marine residual basin can not only show output form of paleo-buried hill and volcanic type, but also can be shown as organic reef, tank and dolomite and weathering crust, etc., with various and complex types. Currently, further research is still lack on a number of master elements, including formation mechanism, mutual relations and spatial distribution, etc. And the problems come from complex surface, complex medium, complex structure and complex reservoir will directly restrict the exploration and development of Pre-Cenozoic marine oil and gas resources. Therefore, in order to push the exploration and development of Pre-Cenozoic marine oil and gas resources forward, the research for the law of Pre-Cenozoic marine deposition distribution, and getting basic knowledge of the formation and evolution of Paleo-Tethys is urgently needed. Under the guidance of activity theory, to make research for finding out the dynamic distribution of lithofacies palaeogeography and tectonic framework of controlling the formation of uplift and depression in basin, especially for finding out the controlling mechanism of the evolution of Paleo-Tethys to marine oil and gas resources of China; through temporal and spatial evolution, making analysis of paleoenvironment; and through the paleoenvironment, making analysis of the distribution of high-energy band around paleocontinental margin (land-sea change and the migration of paleo-shoreline; recovery of lithofacies paleogeography with space-time evolution); forecasting distribution of organic reefs and beaches, and further to determine the space-time distribution of marine stratigraphy in residual basin; giving priority to the study of sedimentation history and space-time development law of marine strata of major basins in China, regional tectonic effect and the control of sea level change on the marine sediment, and the effect of marine sediment to marine oil/gas formation and distribution; further researching for the control of orogenic evolution around margin of basin to the formation and development of basin, and discussing the transformation effect of post-tectonic activities to early marine deposits, and the formation process of a variety of hydrocarbon traps in basin.

Fault system distribution and magmatic activity of residual basin shall be focused on; to mark the distribution of concealed rock and deep fault may be relevant to dolomitisation; to determine the characteristics and activity features of concealed rock and deep fault; to determine the geometry and kinematics characteristics of several tectonic events experienced by Pre-Cenozoic marine deposits; to try to discover the forecast approaches of reservoir and favorable bank distribution through geophysics geotechnical technology.

4.2 Research for oil and gas resources in Pre-Cenozoic marine residual basin

Recently, the development of oil and gas resources exploration in marine residual basin of China has been great. Especially the discovery of Puguang giant gas field in Sichuan confirmed the huge potential of marine oil and gas resources development with firm truth, which shows us prospective future of marine oil and gas resources. However, with the development of exploration, we found the significant breakthroughs in marine oil and gas resources development are mainly located in western and southern China, where a number of problems on surface, medium and reservoir complexity have been hardly to be solved with conventional petroleum geological theory or conventional geophysical exploration means, so new breakthrough and innovations must be fulfilled, so as to be adapted to the characteristics and demand of marine oil and gas research of China.

The Pre-Cenozoic oil and gas resources research is the key field for pushing the development of petroleum geological and geophysical theory and technology. Lower Paleozoic and Upper Proterozoic of China mainly formed marine; Upper Palaeozoic mainly consists of marine-continent cross; Mesozoic and Cenozoic mainly consists of continent, with few marine and deposits. In China, the distribution area of marine strata is around 4.5 million km^2, yet its marine deposit has its own evolution characteristics (with several post-formation, transformation and destroy). Mr. Liu Guangding, the academician, indicated that the dynamic evolution process of "high east and low west" for the Paleozoic Era and "high west and low east" for the Mesozoic and the Cenozoic exists in tectonic frame of China, which is such like a huge seesaw. At this cross band area, marine-continent cross deposits is widely arranged, which should be the key target for searching for marine oil and gas resources[51]. Therefore, recovering paleo-locations in different Eras in researched area, and seeking for high-energy band, including organic reefs, beaches, etc., relevant to paleo-continent margin shall be the significant approaches for Chinese Pre-Cenozoic marine oil and gas exploration.

Huge problems for exploration and development on deep burying, complex and various fault system, a variety types of oil resources, scattered distribution and strong concealment generally exist in oil and gas mineral resources in Pre-Cenozoic marine residual basin. From the innovative research achievements and exploration practices some time ago we can figure out that the oil and gas mineral resources at Pre-Cenozoic marine residual basin can not only show output form of paleo-buried hill and volcanic type, but also can be shown as organic reef, tank and dolomite and weathering crust, etc., with various and complex types. Currently, further research is still lack on a number of master elements, including formation mechanism, mutual relations and spatial distribution, etc. And the problems come from complex surface, complex medium, complex structure and complex reservoir will directly restrict the exploration and development of Pre-Cenozoic marine oil and gas resources. Therefore, in order to push the exploration and development of Pre-Cenozoic marine oil and gas resources forward, the pilot study, including geophysical exploration techniques, shall be made:

1) The study for Pre-Cenozoic marine deposits distribution law, getting basic knowledge of the formation and evolution of Paleo-Tethys; under the guidance of activity theory, to make research for finding out the dynamic distribution of lithofacies palaeogeography and tectonic framework of controlling the formation of uplift and depression in basin.

2) Develop the study of marine oil and gas reservoir formation mechanism, especially the mechanism of dolomitisation. Get knowledge of the complex process of dolomitisation and its transverse change characteristics. Establish reservoir distribution forecast mode in order to instruct the exploration and deployment of oil and gas resources.

3) On basis of further understanding tectonics space-time evolution law and reservoir distribution mode, developing comprehensive geophysical technology applicable for Pre-Cenozoic marine oil and gas prospecting, in order to reveal inside structure, forecast reservoir distribution and mark favorable band.

4.3 The control of global climate evolution to development hydrocarbon-source rocks since the Mesozoic-Cenozoic

The Mesozoic-Cenozoic Era is the times with unique global climate evolution, which get through complete evolution from Hot House environment of the Cenozoic and the Paleogene to Ice House environment of the Neogene. The global temperature in the Cretaceous-Paleogene achieved the highest value since the Cambrian; the global sea level also achieved the highest value in this era, which was 200m and up higher than currently. This high-temperature and high-sea-level era is coincidence with the best development times of global hydrocarbon-source rock, so most of the oil in the world is formed in this times. The coincidence shows that there's obvious relevance between the development of global hydrocarbon-source rock and climate change. The core for the relevance may be that: under the conditions of high temperature and high concentration of atmospheric CO_2 in the Hot House era, terrestrial biomass and ocean productivity was greatly increased, combined with large-scale transgression caused by higher sea level, which formed large area of shallow water shelf deposition, and it strengthened the water layering. With the increasing of organic matter flux, the demand for oxygen is increased accordingly, which enhanced the anoxic environment, so as to made organic matters burying amount increasing, which is favorable for the development of hydrocarbon-source rocks. However, at the regional scale, is there any impact of climate change on the development of hydrocarbon-source rocks? How much of the impact? And what is the mechanism of impact? The aforesaid questions have significant meanings to assessing the development of hydrocarbon-source rocks in specific area.

As an important oil and gas reservoir and origin, the knowledge of hydrocarbon-source rocks development mode of the Mesozoic-Cenozoic in the north China region is significant guidance. In the Mesozoic-Cenozoic Era, with the influence of global climate change, the climate system of China changed from planetary wind system into monsoon climate. With this background, the arid-semi-arid subtropical climate of the north China region controlled by early planetary wind system of the Mesozoic-Cenozoic Era was gradually changed into Neogene monsoon climate control. At the meantime, the tectonic environment was changed from uplift denudation of the end Cretaceous into strong rifting stage of the Paleogene-Neogene, and formed the lower Liaohe–the north China plain rift basin system with basic volcanic eruption at early stage of basin expansion. These major changes must had deep impact on deposition and hydrocarbon-source rocks development, etc. of the Mesozoic-Cenozoic Era in north China region. For example, the major oil-generative assemblage in north China region derived from the Shahejie Formation of the Eocene, which is just coincidence with major global high-temperature and high-sea-level era. Is there any positive relation in between? Whether it means the north China region get through transgression in that era? If no transgression existed, then what's the other major factor? It's a key question. Therefore, another question shall be urgently determined, which is: whether there's impact on the development of hydrocarbon source rock in north China region from global climate and regional climate changes of the Mesozoic-Cenozoic Era? How much the impact is? And what's the mechanism of impact?

Aiming at the problem of direct observation and analysis to deep main hydrocarbon source rock in superimposed basin of China, we shall understand the significance of bio-inversion and sedimentary preservation with the knowledge of paleo-geography and paleo-environment, and shall research the control mechanism to hydrocarbon source rock development from global climate evolution since the Paleozoic Era.

- Reconstructing of stratigraphic framework, paleo-geography and paleo-climate

Through materials collection on basins in the north China region, and comprehensive integration of seismic reflection profile, boring, outcrop and other materials, with other necessary field survey and experimental analysis, the hydrocarbon-source-rock containing system stratigraphic framework of the Mesozoic-Cenozoic Era in the north China region shall be restructured. Through the integrating of regional and global climate changes and paleo-geographic materials, the framework of paleo-geography and paleo-climate of the Mesozoic-Cenozoic Era in the north China region shall be restructured.

- Basin scope tectonics- climate- deposition evolution

Through the sequence stratigraphy analysis to deposition of the Mesozoic-Cenozoic Era in the north China region, focus on the identification of transgression system section; Through the impact analysis on deposition events from tectonics evolution and climate change of basin scope of the Mesozoic-Cenozoic Era, discuss response characteristics of basin deposition record to basin scope tectonics evolution and climate change and geomorphic formation.

- The impact of global climate changes and regional climate-tectonics-paleogeographic structure on the development of hydrocarbon source rock

Through comparison and analysis, to discuss the impact mode, impact degree and impact mechanism of global climate-sea level change and regional climate and paleogeographic structure on the development of hydrocarbon source rock, and to establish a concept mode on basis of relevance between global and regional climate change and hydrocarbon source rock development of the Mesozoic-Cenozoic Era in the north China region, in order to provide base for further researching the impact of global climate change on hydrocarbon source rock development.

4.4 Basin deep fluid-rock interaction and quality reservoir formation mechanism

The basins of China own rich oil and gas resources, but the understanding and knowledge of deep oil and gas accumulation characteristics and distribution law is insufficient. The major area for deep oil and gas accumulation is the low permeability to ultra-low permeability clastic rock and marine carbonate rock system. Research of recent years shows that the oil and gas migration is a geographic process happened along some dominant migration route in complex solid migration channel system formed with cross and interaction among reservoir, fault and unconformity face. The formation of dominant migration route relies on not only the dynamic background of migration, but also the channel environment, including stratigraphic characteristics of the ups and downs, heterogeneity of reservoir, and the development and distribution of faults or fissures. For the study of this complex system, only with qualitative analysis is far more insufficient, so the quantitative basin dynamics study has become the mainstream in this field currently.

The oil and gas resources, as the fluid minerals, its formation, migration, accumulation and destroy and loss after accumulation happed in water-filled rock space (including pores, fissures, caves, etc). The migration and accumulation of oil and gas resources both happened in complex dynamic process of geological history, and it's usually very short in period in comparison with the geological history, so it's hardly to be directly observed in real exploration and study. The oil and gas at all times maintains the trend of flowing in stratigraphic space, and its status and position in geological history depends on the balance relations among the forces applied on it at any time. The heterogeneity of geological conditions and various tectonic activities may complicate the process of oil and gas migration and accumulation, such as severe folding and denudation, lithology of transporting layer and reservoir, spatial changes of physical properties, the separation and connectivity of rupture, etc. Although long term of study has been made, yet people still know very little about the dynamics process of oil and gas migration and accumulation, and its guidance for the work of oil and gas exploration is limited.

In deep environment above the defined economic deadline on oil and natural gas by the conventional "theory of hydrocarbon degradation of late Kerogen", the oil and natural gas still has very high thermodynamic stability. Through research of recent years, we can find that the evolution of oil and gas chemical compositions in thermodynamic balance system of deep basin has no relationship with time, which is mainly controlled by temperature, pressure, and the compositions of reservoir rock and minerals. Deep high-temperature and high-pressure environment, magma activity, the adding of exogenous hydrogen and transition-metal element catalysis will enhance the generation of hydrocarbon and improve the production rate of hydrocarbon. Even for the partial oil-generating I type kerogen material, under favorable deep caprock conditions, it may form gas condensate reservoir with exploration prospects. Through hydrocarbon generation thermodynamic simulation experiment, combining with actual measured results of deep hydrocarbon source rock and oil and gas components, maturity and stable carbon isotope composition, to establish deep hydrocarbon source rock hydrocarbon generation thermodynamic model (kerogen and crude oil relay gas generation model), which is an supplement and improvement for the conventional "theory of hydrocarbon degradation of late kerogen" [84].

With the increase of water/rock and organic/inorganic interaction generated under conditions of burial depth, burial time and deep temperature and pressure, the deep reservoir performs the characteristics of low porosity and low permeability. The development achievements of oil and gas exploration in Jiyang depression have shown that it has two major types of reservoir, including basin deep developed fissure-pore type and pore-fissure type, where the fissures are mainly relevant to (low sequence) faults and abnormal overpressure caprock fracturing; the pores are comprehensive effect results of early "acid dissolution" and deep "alkaline dissolution" and "ultra-pressure support" (sandstone undercompaction) with the background of compaction, so as to break through the bound of the conventional around 3,000m oil and gas industrial base concept, which makes effective reservoir gas-bearing pore rate lowered to 4–5%, and oil-bearing pore rate lowered to 6–7%.

Oil and gas injection is the interaction results between accumulation power and resistance. Different form medium and shallow layers, the increase of deep temperature and pressure of the basin, the physical and chemical properties of hydrocarbon fluid and the interaction of fluid and rocks are changed accordingly. So the mode of oil and gas migration could change, and the accumulation for medium and shallow layers is invalid. In a word, with the increase of burial depth, organic matter tends to pyrolysis, the gas/oil ratio (GOR) is increased, and density of crude oil is decreased, the surface tension of crude oil is decreased, wettability of crude oil is increased, therefore capillary resistance is greatly decreased and the potential of oil injection is improved. To well understand the accumulation condition and character of oil in deep layer, would be useful to enlarge oil and gas development range.

As the results, we need to research into basin deep fluid-rock interaction and formation mechanism of quality reservoir through the study and understanding of basin fluid dynamic evolution mechanism and process, so as to reveal the effect and process of marine carbonate reservoir dolomitization, and to form accurate approaches for forecasting effective reservoir.

4.5 Geophysical response characteristics of oil and gas reservoir and high-precision forecast techniques

Aiming at the problems of complex landform, deep burial depth of target layer, complex structure and complex reservoir physical characteristics of the future explored areas of China, we shall develop geophysical exploration techniques, with the main of seismic technology, supplemented by gravity, magnetic and other field information extraction technology, research and develop software system accurately reflecting geophysical response characteristics of oil and gas reservoir, and establish high-precision geophysical reservoir forecast techniques. In short term, the following key techniques shall be developed:

4.5.1 Complex near surface effect analysis and elimination techniques

Complex near surface effect elimination is the unique problem of Chinese seismic exploration which is different from international study, and it's also the key to break through the exploration of western deep oil and gas resources of China. The research work consists of the following items:

1) Complex surface scattered noise formation mechanism and simulation approaches, which is the foundation for suppressing surface scattered noise.

2) Develop static correction method on basis of 3D tomography, and solve the problems of near-surface velocity modeling and accurate static correction in transverse severe change of velocity and others.

3) Developing the adaptive noise reduction method on basis of noise mode.

4.5.2 Deep non-homogeneous reservoir seismic response characteristics and lighting analysis techniques

To research surface seismic response characteristics of various typical reservoir with deep burial depth and high-impedance shield shall be made, and overlying cap velocity severe change and medium anisotropy shall be taken into account; researching observation system and parameter design method of deep targets in complex overlying conditions. Through lighting analysis to deep exploration targets, the target layer will have wider lighting angle and evener strength research. Owning to severe velocity variation of overlying layer, the current exploration design method based on overlying times can not satisfy the target of deep and complex target exploration, so the new observation system design method on basis of angle lighting must be developed. Only with wider and evener angle lighting, can physical (observed data) foundation be provided for pre-stack inversion and other lithology identification and fluid detection technology.

Lighting Analysis Technology

Through the research of energy distribution (ray density) of underground wave field, the impact of underground structure on the reflected waves can be studied.

Researches include the following items:

1) Developing parallel simulation based on PC cluster applicable for seismic wave simulation in complicated elastic isotropic and anisotropic medium.

2) Geological modeling and surface seismic response characteristics analysis.

3) Observation system design and lighting analysis of deep exploration targets.

4) Empirical model and statistical laws of reservoir rock physical parameter variation and fluid sensitivity analysis techniques in high confining pressure and high temperature, which are used for instructing fluid detection, lithology identification and physical properties inversion.

4.5.3 Basin deep complex construction imaging techniques

Complex surface conditions, severe velocity variation of overlying layer, and long array observation and other elements put forward new requirements for current seismic migration techniques. Therefore, we must develop the following:

1) Pre-stack depth migration and pre-stack time migration method under undulating surface.

Pre-stack Depth Migration

Severe lateral velocity variation has some relations to complex overlying construction, which requires to do depth domain imaging to geological body. The depth migration before superposition of CMP gathers is called the pre-stack depth migration. Establish velocity-depth model, execute depth migration, modify velocity and shape of the reflection interface with interpretation, and with the modified velocity-depth model, repeat depth migration, until the time of depth migration results is equal to the time input into velocity-depth model in depth migration. The output consist of imaging gathers, and the imaging gathers is similar with CMP gathers after time difference correction, but the time coordinate is changed into depth coordinate. However, the imaging gathers consist of migration homing seismic traces, and the superposition of imaging gathers represents depth domain geological body imaging after pre-stack depth migration. If the velocity-depth model with pre-stack depth migration is correct, then the event on the imaging gathers will show the non-difference leveling feature. Two high or too low velocities will cause residual time difference on imaging gathers, and initial velocity-depth model can be modified through analyzing and correcting the residual time difference.

2D pre-stack depth migration has two methods: the first is shot-detection point migration, i.e. to use double square root equation to continue downward with common-shot gather and common receiver gather, and makes the reflected wave energy to be converged on the zero-offect. To retain the zero-offect, and to discard non-zero-offect, the migration profile can be obtained. Another shot gather migration method is based on respective migration of each common-shot gather. In this method, select migrated common-shot gather as common receiver gather, and get sum in each detector point gather, we can get migration profile.

2) Anisotropic pre-stack depth migration and pre-stack time migration, to solve the anisotropic problems due to long array.

3) Preserved amplitude pre-stack depth migration and pre-stack time migration method, and generating initiation angle road gathers techniques, in order to better restore the effective reflection signal of deep target layer, so as to provide quality data for pre-stack inversion.

4) Layer velocity evaluation and time-depth conversion method. Layer velocity evaluation is the basis of velocity migration, and time-depth conversion method has significant meanings to the interpretation and application of time migration results of complex construction.

Pre-stack Time Migration

Time migration before stack of CMP gathers. The reflection is from shot point S to reflected point R, and to receiving point G. The travel time equation of the ray route SRG is as follows:

$$vt = \sqrt{z^2 + (y+h)^2} + \sqrt{z^2 + (y-h)^2} \qquad (1)$$

Where v is medium velocity; t is total travel time from S to R and to G; medium is represented with the coordinate of y (common central point) and z (depth) . Equation (1) can be substituted with the following type:

$$\frac{y^2}{(vt/2)^2} + \frac{z^2}{(vt/2)^2 - h^2} = 1 \qquad (2)$$

Pre-stack time migration can be understood as semi-elliptical scanning stack of equation (1) or diffraction summation of the travel time surface described by equation (2).

Time-depth Conversion

Converse time domain records into depth domain records through velocity.

4.5.4 Deep non-homogeneous reservoir forecast techniques

Deep burial depth and non-homogeneous reservoir make costs and risks of exploration increased, and it put forward higher requirements to reservoir forecast. Therefore, we must develop the following items:

1) Non-homogeneous target reservoir scattering wave imaging technology, which can trap the scattering wave generates in non-homogeneous target reservoir, so as to make aperture reservoir imaging problem being solved easier in the future.

2) High-resolution non-linear wave impedance and physical parameter inversion techniques.

3) Angle domain pre-stack inversion of elastic parameters, to address the lithology and fluid identification detection key issues.

4) Seismic attribute analysis and sub-consciousness techniques with fluid type.

4.5.5 Multi-wave multi-component seismic methods and key techniques

Multi-wave seismic exploration technology has significant meanings to imaging under high-resistance shield layer, identification fissure distribution and movement and reservoir fluid properties detections. On basis of inversion of multi-wave information, more physical parameters can be obtained. Therefore, the research for multi-wave information processing and imaging techniques shall be developed, including:

1) Multi-wave information noise suppression and converted wave static correction techniques.

2) Converted wave pre-stack time migration techniques.

Converted Wave

When the wave, either longitudinal or transverse wave, tilts to the elastic interface, the reflection transverse wave, reflection longitudinal wave, transmission transverse wave and transmission longitudinal wave can generate. The wave which is same with the incident wave type is called the same wave, while the wave whose type is changed is called converted waves. Converted-wave reflection and transmission follow the Snare Law. Energy of converted wave has relation to incident angle, and vertical incidence can not form the converted wave; only when the incident angle is large enough, sufficient energy converted waves are recorded. Therefore, in seismic exploration, mainly the same wave is used. Only in some special issues have the converted wave been used, such as in thin layer studies, utilizing the converted shear wave, the resolution can be increased [86].

---

3) Multi-component pre-stack depth migration techniques.

4) Fissure identification and anisotropic parameter extraction techniques.

4.5.6 Deep reservoir weak seismic signal processing key techniques

As the complex surface conditions, deep burial depth, high impedance shield, strong noise interference at deep exploration surface and weak effective reflected signal problems, therefore targeted signal processing technology must be studied, including:

1) Pre-stack noise suppression techniques.

2) Layer multiple attenuation techniques.

3) Effective signal frequency band recovery and frequency improvement techniques.

4) Rules access, regularization method of evenly covered pre-stack data.

5) Dynamic correction stretch recovery techniques aiming at long array.

4.5.7 Deep structure and reservoir distribution of the bit field information extraction technology

Gravity, magnetic and electric and other non-seismic methods have significant application values for marking magma and fault zone, stripping the substrate effect, and directly marking oil and gas favorable area and volcanic oil and gas reservoir. The following items will be researched:

1) Pre-Cenozoic residual basin remaining thickness and volcanic reservoir bit field extraction techniques.

2) Fault and volcanic reservoir bit field extraction techniques relevant to deep oil and gas.

3) Actual gravity and magnetic material processing and interpretation technical process flow of deep oil and gas exploration target area.

4.5.8 Processing process flow and technical integration of deep oil and gas exploration

Taking the above-mentioned key techniques as the core, to form a set of seismic and non seismic material processing process flow and quality control system for deep marine reservoir, deep continent clastic reservoir and volcanic reservoir oil and gas exploration, and to integrate a set of key technique and software module series aiming at oil and gas exploration for various special geographic conditions.

4.6 Techniques for enhancing oil and gas recovery

The major cause for low recovery of primary oil extraction is the depletion of reservoir energy, thus the method of manual water and gas injection for adding reservoir energy, maintaining reservoir pressure for secondary oil extraction is the major development type used all around the world. However, after secondary oil extraction, there's still 60–70% of remaining crude oil can not be mined underground, so the different tertiary recovery methods have emerged.

Currently, the have-been-tried tertiary recovery methods mainly include chemical method, miscible method, thermal method and microbiological method, etc. According to different action principles, the chemical method can also further divided into alkaline enhancing, polymer enhancing, surfactant enhancing, and the developed Alkali-Polymer enhancing (AP enhancing), Alkaline-Surfactant-Polymer enhancing (ASP enhancing) or Surfactant-Alkali-

Polymer enhancing (SAP enhancing), etc. The miscible method is further divided into solvent miscible enhancing, hydrocarbon miscible enhancing, CO_2 miscible enhancing, N_2 miscible enhancing, and other inert gas miscible enhancing. In recent years, a gas-water alternate enhancing (WAG enhancing) has been developed. Thermodynamic oil recovery method includes steam enhancing, burning reservoir, etc. The above-mentioned existing oil recovery technology usually can achieve 30% to 40% of the recovery. That means 60% to 70% of the crude oil remain in the crude oil reserves is not extracted in existing geographic reserves. Therefore, searching for new mining technology for improving the recovery has a very important significance to solving the problem for oil shortage to China.

The current developed and researched latest tertiary recovery technology is the microbial enhanced oil recovery technology (referred to as MEOR). The microbial enhanced oil recovery technology has the features of simple construction, not damaging reservoir, not affecting crude oil quality, pollution-free, less investment, quick recovery, and high benefits. It can improve the degree of reservoir heterogeneity and improve oil recovery to a large extent. Taking it as enhancing system is an innovation, which can better adapt to reservoir characteristics of high water cut stage. It is a new technique combining profile control and displacement together. It mainly refers to ground isolation and culturing of micro-organisms bacilli and filling nutrient solution into the reservoir (exogenous microbial enhanced oil recovery), or simply filling the nutrient solution into the reservoir to activate the endogenous micro-organisms and make them growing and reproducing in the reservoir (endogenous microbial enhanced oil recovery), so as to enhance oil recovery. Microbial enhanced oil recovery effect principles can be divided into 3 types: the first is to degrade high molecular weight compounds into low molecular weight compounds through microorganisms, so as to reduce the viscosity, in order to facilitate the exploitation of crude oil; the second is to produce a large number of CO_2 and CH_4 gas through the role of microorganisms, thereby increasing the reservoir pressure, in order to facilitate the exploitation of crude oil; the third is to produce surface-active agent through the role of microorganisms, to reduce the surface tension of oil layer, in order to facilitate the output of crude oil.

At present, oil field microbial technology mainly includes two aspects of the research content: one is the technology on exogenous microbial enhanced dense oil recovery; the other is the technology on endogenous microbial enhanced oil recovery. Owing to high cost and low recovery, the exogenous microbial enhanced dense oil recovery technique has gradually substituted by endogenous microbial enhanced method, such as the Yumen Field in west China region. With developing of field extraction, the research institute of the Field adopted adding imported exogenous bacterium to enhance dense oil recovery, but effect is little, which is mainly because acquisition process is too long, bacteria survive in harsh conditions, and the survival of exogenous bacteria is not known. Therefore, in view of recovery was little improved and high cost, they gave up exogenous bacteria for improving heavy oil recovery technology. The endogenous bacteria for improving oil recovery technology is low in cost, and mature in technology, and activation process is easy to implement. More superior is the micro-organisms can also be used to block the high permeability channel of the reservoir. After many years of water injection, the injected water may bypass the seepage high-resistance oil-containing area, and flow along the channel with minimum seepage resistance. The number of micro-organisms is also a lot in this channel. Endogenous bacteria for improving oil recovery technology can activate the endogenous micro-organisms by adding nutrients into injected water. The propagation of micro-organisms causes its surge in number, in order to block invalid waterways, expense wave and volume, and improve water injection efficiency. Therefore, the technology of endogenous bacteria for improving oil recovery has got more and more attentions.

Therefore, the main thrust of Chinese Academy of Sciences should be: to have an accurate understanding of the distribution characteristics and the formation mechanism of remaining oil in highly developed oil-containing gas reservoir, to research and develop new tertiary oil recovery displacing agent on basis of micro-organisms, chemical enhancing, nano-materials and others; to adopt an integrated approach to improve the technology of oil and gas recovery.

4.7 High-density and wireless transmitted seismic acquisition system on basis of MEMS sensor

The existing oil and gas geophysical exploration technology is formed in exploration practices of basin shallow exploration, which is only applies to oil and gas exploration with simple surface conditions, construction and reservoir. In order to meet the needs for basin oil and gas exploration, the deep oil and gas detection technology with complex surface, deep buried depth, complex structure, strong reservoir heterogeneity and other features is urgently needed. To research and develop a new generation of high-precision digital seismic data acquisition system for deep oil and gas exploration, so as to provide crucial technical support for the deep oil and gas exploration and development of basins. The MEMS-based seismic data acquisition system shows extremely broad prospects for development. To master the core technologies with independent intellectual property rights to form a high-volume production capacity is essential to high-precision deep seismic exploration of China. So we should focus on:

4.7.1 The development of MEMS digital geophone

The development of MEMS digital geophone includes three aspects: the MEMS acceleration sensor device, capacitance-voltage conversion circuit and an integrated digital detector.

4.7.2 Key technologies of seismic data acquisition system

- Efficient and high-speed data transmission technology
- Wired-wireless data transmission, wireless data transmission and protocol

To research wireless transmission technology between base stations and collection stations; through connection between MCU interface and control nodes on main line, to realize wireless and wired information cross-transmission of seismic data acquisition system.

- 10,000-scale digital seismic instruments

The key is to solve the problems of equipment cost, equipment power consumption, equipment reliability and equipment weight, etc. Focus on distributed data transmission and control system with flexible expansion structure and parallel processing.

4.7.3 Production process quality monitoring platform of MEMS detector

- Control and monitoring of sensing device fabrication and packing process

To meet the high-volume demand for MEMS acceleration sensor devices by oil exploration equipments, and to strengthen development studies from science and technology platform to the production platform of sensor devices research and development, the research, development and building of sensing devices control and monitoring platform shall be a key.

- μg-level sensing device vibration test

The test platform, which includes ultra-quiet environment and high-precision vibration tester, is expected to become μg-level vibration standard test platform for sensing devices after building.

4.7.4 The formation of productive capacity of seismic data acquisition system

To be actively involved in the revolutions of geophysical exploration ideas and methods, to be self-reliance to develop core technology of MEMS sensors, in order to form a high-density and wireless-transmission seismic acquisition system, and to built the corresponding capacity.

4.8 Parallel computing technology of massive data processing and simulation of large-scale oil and gas reservoir

To integrate resources and to establish a massive data processing grid computing technology; to develop parallel algorithm software system being capable of executing large-scale numerical simulation of porous media multiphase fluid.

To research and develop parallel computing technology for mass data processing and large-scale oil and gas reservoir simulation on basis of parallel processing accelerator. The parallel processing accelerator uses a standard PCI card, which can be installed on the server. As its main functional chip, the parallel processing accelerator FPGA (programmable gate array) and DSP (digital signal processor) is used to process mass data, and introduce DSP function in the system, so as to improve acceleration performance.

This accelerator consists of hardware and software, where the hardware refers to PCI card, which has powerful processor chips; the software refers to the development environment and logic algorithms.

The realization of the function of parallel processing accelerator mainly relies on FPGA and DSP.

FPGA is the programmable logic gate array, which can realize the function as coprocessor. FPGA can execute data pre-processing to reduce the burden on CPU, and to play the advantages of CPU serial computing.

Via making the inherent parallel computation part in high-performance computing algorithms being hardwared, the HPC application acceleration can be achieved. In fact, the parallel computation we frequent mentioned can be divided into multiple levels: level I: to execute task multi-threaded allocation in a number of CPU in the cluster computing, which we can call the "task-level parallelism"; level II: we can call it the "instruction parallelism". The conventional CPU supports a limited number of instruction concurrent processing, i.e. the CPU instruction pipeline amount or the released amount is limited. However, the FPGA can provide many pipelines, i.e. it can execute a large number of parallel instructions at the same time. The "data parallelism" is the third-level parallel processing capability that the FPGA is very easy to achieve. The structure of FPGA makes it very easy to implement parallel operation. Thus, through the configuration, it can simultaneously operate a large number of data throughputs. In this case, the device is equivalent to the operation of several conventional CPUs at the same time.

5 The Construction of Chinese Oil and Gas Science and Technology Innovation System

The development of oil and gas technology must be judged by framework of the national macro-economic development strategic planning, so as to get sufficient attention and protection. So it would be necessary to draw on foreign experience on the basis of the above work, in order to build the oil and gas technology innovation system suitable to China.

5.1 General foreign policies and measures

As the uneven distribution of the global oil and gas resources, it leads to differences in national oil and gas resource endowments, thus the strategies also vary considerably in the protection of oil and gas supply and development of oil and gas technology among states. By oil and gas self-sufficiency level, the states can be divided into three categories: oil and gas net exporter, major importer and producer of oil and gas, and oil and gas net importer, where:

1) The strategies of the oil and gas net exporter are: to establish energy depletion strategies, to develop oil and gas technical services, in order to achieve the sustainable development in case of oil and gas depletion; to ensure oil and gas industry owing sufficient investment and new reserves, to enhance the balance development of domestic oil industry; to make oil and gas prices stable through various means mainly including dialogue; to encourage private capital participation in oil and gas infrastructure construction; to be concerned about the exploration of unconventional oil and gas resources. This category refers to Norway and Canada, etc.

2) The strategies of major importer and producer of oil and gas are: to eliminate the barriers of oil and gas exploration, in order to increase domestic oil and gas reserves and production; to adopt new technologies to increase the output of old oil and gas wells; to encourage oil and gas infrastructure; to promote oil and gas market liberalization; to establish national strategic petroleum reserves; and to develop alternative energy sources, etc. United States

is a typical representative of such countries.

3) The strategies of major exporter and producer of oil and gas are: to encourage their domestic companies in overseas oil and gas exploration and development; to promote the liberalization of oil products in the domestic markets; through the tariff structure to protect the competitiveness of the domestic refining industry; to strengthen unconventional oil and gas resources (such as natural gas hydrate) exploration and development; to promote the development and use of alternative energy sources. Japan and South Korea are typical representatives of such countries.

5.2 Improving the oil and gas science and technology innovation system and improving the oil and gas science and technology capability of independent innovation

The oil and gas science and technology innovation system consists of three major subsystems: the knowledge creation systems, technology development system and technology intermediary service system, where the research institutes and universities engage in scientific and technological energy knowledge production activities, which are the main body and a base for knowledge innovation. The energy enterprises are the main body of technological innovation, which under the inspiration by the market mechanism, have driving force and ability of technological innovation. They are the main body of technological innovations investment, technological innovation activities, and benefits and risks commitment. The energy intermediary service agencies are the bridge and link to communicate the energy technology innovation, technology flows and transformation. The subsystems all include fund, human resources, institutions and other innovation resources. The target of energy technology innovation system building is pushed forward from scientific and technological development to technology industry.

1) Strengthening the building of national oil and gas science and technology innovation system.

To establish national oil and gas scientific and technological innovation system combining production and research together; to give full play to the complementary strengths of national oil companies, universities and national research institutes, to substantially increase independent innovation ability of oil and gas technologies of China. The oil companies should increase fund investment and form research and development institutions with their advantages of strong financial background and extensive field data, and attract universities, national research institutions and other technological innovations to carry out cooperative researches, in order to form the main body of oil and gas technology innovation. The colleges and universities should actively participate in the oil and gas technology research and development with their advantages of wide-covered disciplines and firm theoretical research foundation, etc., in order to try to become an important force for innovation in basic theory. As a national scientific research institution, the Chinese Academy of Sciences shall be the key component of oil and gas science and technology innovation and major supplier of related equipments and technical services. Therefore, Chinese Academy of Sciences should be focused on building a base for oil and gas science and technology to improve oil and gas science and technology independent innovation capacity. It shall try to become the main strength of the original scientific innovation and major supplier of related equipments and technical services.

2) Strengthening basic research, and strengthening science and technology systematic layout of the future oil and gas.

Oil and gas technology has features of large investment, long term and high risks, etc. Once breakthrough is made, it would have continuing advantages and path dependence. Therefore, we must strengthen the systematic layout. Basic research is the foundation to maintain continuous innovation and independent innovation capability, so forward deployment and long-term sustainability is a must.

3) Breaking through key technologies in oil and gas exploration and development and realizing great-leap-forward development.

China has complex geological conditions and great potential in oil and gas, so aiming at important scientific and technological issues with a certain foundation and advantages for critical and strategic importance, studies shall be made, in order to gain achieve leapfrog development of petroleum science and technology.

4) To emphasis on appropriate technology integration of high-tech and oil and gas exploration and development to enhance industrial capacity.

The changes from focusing on individual technology research to the industry-centric integration of major products and industries, the changes from market-for-technology to the introduction, digestion and absorption of re-innovation, and the changes from the technical indicators results-oriented to the market-oriented and industrialization-oriented shall be realized, so as to achieve new breakthrough in oil and gas innovation in key areas relevant oil and gas resources recovery improvement.

5) Emphasizing on unconventional oil and gas resources exploration and development.

China is rich in natural resources of unconventional oil and gas, but due to technical and economic reasons, the overall level of exploration and development is low, most are still in the survey and pre-investigation, or the initial development stage. However, in the future, along with the continued development of conventional oil and gas resources and consumption, non-conventional oil and gas resources will become the main body of the future development of oil and gas resources. Thus, Chinese unconventional oil and gas resources are yet to be developed areas. The country should put forward the appropriate policies or incentives to encourage unconventional petroleum resources exploration and development technology research and exploration practices, including: to establish unconventional oil and gas resources development technology fund through financial subsidies, in order to enhance scientific research efforts; in the early development of unconventional oil and gas resources, implement tax and income tax relief policy, to promote the exploration and utilization of unconventional oil and gas resources to form an important complement to conventional oil resources.

6) Increasing oil and gas science and technology investment, ensuring getting stable support for oil and gas technology research and development.

Due to the high cost of oil and gas innovation and technology research and development, high technological risks and long rewards cycle, it usually can not form business competitiveness, so it lacks investment attractiveness of enterprises. Therefore, the government should be the most important capital investment in such technologies. On the basis of government investment, the technology still relies on enterprise to make it transferred into commercial gain to realize long-term and continuous technology development target. Thus petroleum companies need to be incorporated into the development framework. Therefore, oil and gas technology development requires sustained and stable investment, in particular by strengthening the original oil and gas technology, innovation and basic scientific research, cutting-edge technology research, as well as the basic technology conditions investment and support; at the same time, to enhance the efficient use of oil and gas science and technology investment funds, and to establish and improve the science and technology funds management system.

7) To strengthen multi-level human resources training, and to build a stable team of oil and gas science and technology innovation.

Scientific and technological innovation relies on human resources. Human resources have become the most important strategic resources. So we shall implement the energy technological innovation HR strategy, and actually strengthen the building of energy science and technology personnel, in order to provide personnel protection for the implementation of energy technology innovation and security system. To increase the intensity of the cultivation of academic leaders relying on major projects and scientific research base, as well as international academic exchange and cooperation projects, so as to actively promote innovative team building. To focus on discovering and developing a group of young high-level experts; to enhance organic combination of energy technological innovation and personnel training, to support the participation of scientific and technological innovation by professional post-graduates, in order to enhance their interests and the scientific spirit of innovation and practice in science and technology innovation practices.

8) Strengthening science and technology policy guide, to promote the rapid development of oil and gas science and technology.

Oil is the main energy, and petroleum science and technology development, to a certain extent, on behalf of the level of energy technologies of China. Oil and gas science and technology inputs, oil and gas science and technology personnel construction, and improvement of oil and gas science and technology system all rely on unified guidance of national policies with long-term endeavors, in order to provide technological support to sustainable and stable development of China. The Chinese Academy of Sciences shall play a role as leader, model and backbone, and strengthen policy orientation, so as to make its due contributions to the rapid development of Chinese oil and gas technology.

References

[1] Yang Lei. The international oil price 147 U.S. dollars a barrel intraday for the first time. http:// news.xinhuanet.com/fortune/2008-07/12/content_8532842.htm. 2008-07-12.

[2] Campbell C J, Laherrère J H. The end of cheap oil. Scientific American, 1998, March:78-83.

[3] Chinese Academy of Engineering. Strategic Research Report on Sustainable Development of Oil and Gas Resources of China.2004.

[4] IEA. World Energy Outlook 2008. Organisation for Economic Co-operation and Development, International Energy Agency, Paris, 2008.

[5] EIA. Annual Energy Outlook 2008 with Projections to 2030. Energy Information Administration, Washington DC,2008.

[6] IEA. World Energy Outlook 2002. Organisation for Economic Co-operation and Development, International Energy Agency, Paris, 2002.

[7] IEA. Energy Technology Perspectives. Organisation for Economic Co-operation and Development, International Energy Agency, Paris and Washington, DC, 2008.

[8] EIA. International Energy Outlook 2007. Energy Information Administration, Washington DC, 2007.

[9] Aleklett K. Peak Oil and Evolving Strategies of Oil Importing and Exporting Countries: Facing the Hard Truth about an import decline for the OECD Countries. Discussion Paper No.2007-17. Dec.2007.

[10] EIA. Annual Energy Outlook 2007 with Projections to 2030. Energy Information Administration, Washington DC,2007.

[11] USGS. US World Geological Survey Petroleum Assessment-Description and Results, Washington, DC. 2000.

[12] BP. Statistical Review of World Energy 2009.2009.

[13] Anon. Worldwide look at reserves and production. Oil & Gas Journal,2008,106(48):22-23.

[14] Liu Zengjie. 2007 world oil and gas reserves, production analysis. http://www.lrn.cn/zjtg/academicpaper/200803/t20080328_213434.htm.

[15] Ministry of Land and Resources,National Development and Reform Commission, Ministry of Finance. Introduction of the new national oil and gas resources assessment results.2009.

[16] Hubbert M K. Nuclear Energy and the Fossil Fuels. Shell Development Company,Huston,Texas,1956.

[17] Chen Xirong. Oil civilization chronology（from 10th century BC to 1997）.Oil Forum,1999,（4）:86-90.

[18] Hubbert M K. Degree of advancement of petroleum exploration in the United States. AAPG Bulletin,1967,52(11)：2207-2227.

[19] Brown L R. Beyond the oil peak. In: Plan B 2.0: Rescuing a Planet Under Stress and a Civilization in Trouble New York. 2003.

[20] IEA. World Energy Outlook 2006. Organisation for Economic Co-operation and Development, International Energy Agency, Paris, 2006.

[21] China Petrochemical Corporation (Sinopec). Annual Report 2002.2003.

[22] Gong Xiang, Huang Gangwei. China's oil and natural gas strategic resource analysis. http://energy. amr.gov.cn/edsoil/ViewArticle.do?articleid=3976. 2007-02-13.

[23] Wu Rongqing. 50% of the oil recovery is not far off. http: //www.lrn.cn/zjtg/ societyDiscussion/200907/ t20090714_389514.htm.

[24] Zhao Wenzhi, Zou Caineng, Song Yan, et al. Progress of Petroleum Geological Theory and Method. Beijing：Petroleum Industry Press，2006.

[25] Liu Guangding. The second exploitation of the oil and gas resources in China. Progress in Geophysics，2001，16（4）：1-3.

[26] AAPG. 2006 Technical Program. 2006. http://aapg.confex.com/aapg/2006am/index.epl. 2007-12-20.

[27] AAPG. 2008 Technical Program. 2008. http://www.aapg.org/sanantonio/tech_program/sessions. cfm.2008-05-16.

[28] Wu Zhengquan, Song Jianguo. The evolution of world oil and gas exploration strategy and the inspiration of China's onshore oil and gas exploration. Oil Forum,2003,(10):32-41.

[29] Andrews E B . Rock Oil, its geological relations and distribution. The American Journal of Science and Arts, 1861,32(94):85-93.

[30] White I C. The geology of natural gas. Science, 1885,5(125):521-522.

[31] Hager D.Practical Oil Geology: The Application of Geology to Oil Field Problems. McGraw Hill,1915.

[32] Liu Wenlin,Zhang Yin. Review, inspiration and suggestions of petroleum geophysical development history. Oil Forum,2003, (10): 42-52.

[33] Levorsen A I. Geology of Petroleum. San Francisco:Freeman,1954.

[34] Wu Fengming. Petroleum geology review and prospect of hundred years of history:From 1859, Derek, "the world's first well" talk about. Oil Forum, 1999, (5):65-71.

[35] Wegener A. The Origin of Continents and Oceans. Brunswick: Vieeweg,1915.

[36] Hess H H. History of ocean basins. Petrologic studies—A volume in honor of A. F. Buddington. Geol. Soc. America. 1962: 599-620.

[37] Research Group for Earth Science Development Strategy of Chinese Academy of Sciences. Strategic Report: China's Earth Science Development for 21Century. Beijing :Science Press,2009.

[38] Tissot B P.Premiers donnees surles mecanismes et la cinetique de la formation du petrole dans les sediments: simulation d'un schema reactionnel sur ordinateur. Revue de Institut Francais du Petrole, 1969, 24 (5):470-501.

[39] Liu Zhenwu, Fang Zhaoliang. Progress in Petroleum Technology. Beijing :Petroleum Industry Press,2006.

[40] Wei Yiming, Fang Zhaoliang, Li Jingming,et al. Report on Chinese Upstream Oil Industry and Technology Policy Research. Beijing :Science Press,2006.

[41] Magoon L B. Dow W G. The Petroleum System: From Source to Trap. AAPG Memoir 60,1994.

[42] Zhao Wenzhi, He Dengfa. Petroleum system theory in the oil and gas exploration. Explorers,1996, 1(2):12-19.

[43] Wang Hua. The basic principles, method and application of sequence stratigraphy. Beijing:China University of Geosciences Press,2008.

[44] BP. Statistical Review of World Energy 2008.2008.

[45] Liu Guangding, Zhu Qianyi. Latest proceeding of petroleum geophysics. Progress in Geophysics, 2003,18(3):363-367.

[46] Fang Zhaoliang,Wu Mingde,Feng Qining. Key logging technology outlook. Oil Forum,2005,(2):32-35.

[47] Li Chuanle, Wang Anshi,Li Wenkui. Summarising and analyzing about " explode in formation" using in oil and gas well of overseas. Oil Drilling & Production Technology,2001,(6):77-78.

[48] Van Hamme J D, Singh A, Ward. O P. Recent advances in petroleum microbiology. Microbiology and Molecular Biology Reviews,2003,67(4):503-549.

[49] Pan C H. Nonmarine origin of petroleum in north Shensi, and the Cretaceous of Szechuan, China. AAPG Bulletin,1941, 25: 2058-2068.

[50] Liu Guangding. The progress of earth sciences in 20th century.Progress in Geophysics, 2000,15(2):1-6.

[51] Liu Guangding. The work of mineral resources is in command by the scientific development view. Land and Resources,2006,(1):4-7.

[52] He Yanqing. Scientific and technological progress to promote the sustainable development of the world oil industry. Oil Forum,2006,(6):28-31.

[53] Liu Zengjie. Brazil Oil and Gas Resources Situation.2009. http://www.lrn.cn/invest/

internationalres/200904/t20090422_357178.htm.

[54] Anon. Production goes on stream in Tupi: year I of a new era. http://www2.petrobras.com.br/portal/frame.asp?area=apetrobras&lang=en&pagina=/Petrobras/ingles/area_tupi.asp. 2009-07-12.

[55] Anon. Technology drives today's petroleum industry. Oil & Gas Journal,2006,106(3):4-6.

[56] Shell. Shell Technology Report: The Power of Innovation. 2007.

[57] About Schlumberger. http://www.slb.com/content/about/index.asp?

[58] USGS.Circum-Arctic Resource Appraisal: Estimates of Undiscovered Oil and Gas North of the Arctic Circle.2008.

[59] DOE. An Interagency Roadmap for Methane Hydrate Research and Development.2006.

[60] Yamazaki A.MITI's Plan of R & D for Technology of Methane Hydrate Development as Domestic Gas Resources. 20th World Gas Conference.1997.

[61] Expert Committee of CAIL,ONGC, and DGH. National Gas Hydrate Program. 1996.

[62] USGS. Assessment of Gas Hydrate Resources on the North Slope, Alaska. Fact Sheet 2008-3073.2008.

[63] ExxonMobil. Taking on the World's Toughest Energy Challenges.2007.

[64] Jia Chengzao, Zhao Wenzhi, Zou Caineng,et al. Lithologic stratigraphic reservoirs geological theory and exploration techniques. Beijing :Petroleum Industry Press,2008:1-346.

[65] Spencer C W. Review of characteristics of low-permeability gas reservoirs in Western United States. AAPG Bulletin, 1989, 73(5) : 613 - 629

[66] Zhao Chenglin. Reservoir Sedimentology. Beijing :Petroleum Industry Press,1998.

[67] Li Daopin,Luo Diqiang,Liu Yufen,et al. Low-Permeability Sandstone Oil Field Development. Beijing :Petroleum Industry Press,1997.

[68] EC. Prospective Analysis of the Potential Non-conventional World Oil Supply: Tar Sands, Oil Shales and Non-conventional Liquid Fuels from Coal and Gas.2005.

[69] Task Force on Strategic Unconventional Fuels. Development of America's Strategic Unconventional Resources.2006.

[70] Zhang Jie,Li Bing. Unconventional oil and gas resources is moving toward industrialization. China Petroleum and Chemical Industry,2008,(7):26-28.

[71] Salvador A.Energy:A Historical Perspective and 21st Century forecast.Zhao Zhenzhang,Hu Suyun,Li Xiaodi,et al. Trans. Beijing :Petroleum Industry Press,2007.

[72] Liu Honglin, Liu Hongjian,Li Guizhong,et al. how about coalbed methane status in China energy industry and its utilization foreground. China Mining Magazine ,2004,13(9):11-15.

[73] WEC. Survey of Energy Resources 2007. http://www.worldenergy.org/documents/ser2007_final_online_version_1.pdf.

[74] Stephen A H. Tight gas sands. SPE Journal, 2006, (1):86-93.

[75] Guan Deshi. Unconventional Oil and Gas Geology in China. Beijing :Petroleum Industry Press,1995.

[76] Dong Yunlong. Low-permeability oil and gas resources will be mainstream in the development of China's future. http://news.cnpc.com.cn/system/2009/03/26/001230359.shtml. 2009-03-31.

[77] Curtis J B.Fractured shale-gas systems.AAPG Bulletin,2002,86(11):1921-1938.

[78] Richard M P. Total petrolenm system assessment of undiscovered resources in the giant Barnett Shale continuous (unconventional) gas accumulation, Fort Worth Basin , Texas. AAPG Bulletin,2007 ,91 (4) :551-578.

[79] Zhang Jinchuan, Xu Bo, Nie Haikuan,et al., Exploration potential of shale gas resources in China. Natural Gas Industry, 2008, 28 (6) : 136 - 140.

[80] Zhang Zishu.On the water soluble gas. Natural Gas Geoscience,1995,6(5):29-34.

[81] Wu Xiaochun, Pang Xiongqi, Yu Xinghe, et al. Discussion on main control factors and evaluation methods in the concentration of water soluble gas. Natural Gas Geoscience,2003,14(5):416-421.

[82] Fang Zhaoliang,Liu Wenlin, Wang Shangxu. Prospects of key technologies on geophysical

exploration. Oil Forum,2005,(4),4-10.

[83] Wen Baihong, Yang Hui, Zhang Yan. Non-seismic techniques for oil and gas exploration in China. Petroleum Exploration an Development,2005,32(2):68-71.

[84] Tissot B,Pelet R. Nouvelles donnees surles mecanismes de genese et de migration du petrole:simulation mathematique et application a la prospection. Proceeding 8th World Petroleum Congress,1971:35-46.

[85] Oz Yilmaz . Seismic Analysis : Processing, Inversion and Interpretation of Seismic Data.Liu Huaishan,Wang Kebin,Dong Siyou,et al.Trans.Beijing :Petroleum Industry Press,2006.

[86] Anon. Converted wave. Chinese Geophysical Information Network. http://www.wutan.cn/html/ zhishiku/wutancidian/dizhen/200804/18-2182.html

Epilogue

Based on the research of development of world science and technology, focusing on solving major technological issues of China in development, the Chinese Academy of Sciences made deployments on scientific and technological research work in a number of important scientific and technological fields for the 2050, which has significance to the scientific and technological development of China in the future and forward-looking. Since the startup of Technological Development Roadmap for Oil and Gas Resources to 2050 in China, under the joint efforts of the study team members for more than a year, with the instructions of leaders from Chinese Academy of Sciences, the Strategic Planning Council, Resources and Environment Science and Technology Agency, and etc., supported by many experts both inside and outside the Academy, with reference to a number of research achievements on oil and gas technology strategic planning by relevant units, we completed the "Oil and Gas Resources in China: A Roadmap to 2050".

The report systematically elaborates on the demand and supply situation of Chinese oil and gas resources, analyzes the development status and future development trends of domestic and international oil and gas resources technology, puts forward the short-term (by 2025), medium-term (by 2035) and long-term (by 2050) development goals of China, and proposes ways to achieve these goals. Moreover, aiming at the layout of Chinese oil and gas scientific and technological research institutions, research fields and development orientation of Chinese Academy of Sciences, the report as well puts forward the main research orientations by Chinese Academy of Sciences in the field of oil and gas resources of China in the future.

Owing to unpredictability of future development and limited academic level of the researchers, the report is inevitably defective. Comments are welcome.

Research Group on Oil and Gas Resources of the Chinese Academy of Sciences

May, 2010